T0297361

CONCEPTUAL BREAKTHROUGHS IN EVOLUTIONARY ECOLOGY

Comments from the Series Editor, John C. Avise

The "Conceptual Breakthroughs" (CB) series of books by Elsevier aims to provide panoramic overviews of various scientific fields by encapsulating and rating each discipline's major historical achievements in an illuminating chronological format. Each volume in the CB series is authored by a world-leading expert who offers his or her personal insights on the major conceptual breakthroughs that have propelled a field forward to its current state of understanding. Intended for advanced undergraduates, graduate students, professionals, and interested laypersons, the dozens of essays in each CB book recount how and when a recognizable discipline achieved major advances along its developmental pathway, thereby offering readers a pithy historical account of how that field came to be what it is today.

This is the third volume in the CB series—all written in the same concise style and format—and all intended for an intellectually curious audience ranging from laypersons and beginning students to advanced practitioners. The first two books in the CB series were as follows:

Conceptual Breakthroughs in Evolutionary Genetics by John C. Avise (2014).

Conceptual Breakthroughs in Ethology and Animal Behavior by Michael D. Breed (2017).

CONCEPTUAL BREAKTHROUGHS IN EVOLUTIONARY ECOLOGY

LAURENCE MUELLER

Professor, Department of Ecology and Evolutionary Biology, University of California, Irvine, CA, United States

Academic Press is an imprint of Elsevier
125 London Wall, London EC2Y 5AS, United Kingdom
525 B Street, Suite 1650, San Diego, CA 92101, United States
50 Hampshire Street, 5th Floor, Cambridge, MA 02139, United States
The Boulevard, Langford Lane, Kidlington, Oxford OX5 1GB, United Kingdom

Library of Congress Cataloging-in-Publication Data
A catalog record for this book is available from the Library of Congress

British Library Cataloguing-in-Publication Data
A catalogue record for this book is available from the British Library

ISBN: 978-0-12-816013-8

For information on all Academic Press publications visit our website at https://www.elsevier.com/books-and-journals

Publisher: Charlotte Cockle
Acquisition Editor: Anna Valutkevich
Editorial Project Manager: Andrea Dulberger
Production Project Manager: Swapna Srinivasan
Cover Designer: Christian J. Bilbow

Typeset by TNQ Technologies

To my wonderful wife, Carol Joy Krieks, and daughters,
Adrienne and Aellana.

Contents

Introduction

The field of evolutionary ecology is relatively new. Of course, the important role of an organism's ecology in the outcome of evolutionary change was recognized by Darwin (Chapter 1). But the modern development of evolutionary ecology as a separate field is much more recent. For me a seminal paper that set in-motion the development of evolutionary ecology as an important discipline was MacArthur's 1962 paper (Chapter 18) on density-dependent population growth and fitness. Certainly, by the 1970's the field was in full bloom. In 1987 the journal *Evolutionary Ecology* published its first edition.

Identifying research that belongs exclusively to evolutionary ecology as opposed to simply evolution or ecology requires at some point a subjective analysis. In this book I have primarily focused on research that has well defined elements of both ecology and evolution. However, as a means of background development some chapters that are basically ecological in flavor (Chapters 3 and 8) and some others that are largely evolutionary (Chapter 4).

The subjective meter is elevated another notch when deciding what constitutes a breakthrough. Citations are certainly a useful objective measure but are affected by research intensity in different fields. Review articles are often heavily cited but don't necessarily contain new results that would be considered breakthroughs. However, there are exceptions like Stearn's very thoughtful review and synthesis of life-history research in 1977 (Chapter 41). Some papers like Norton's work on selection in age-structured populations (Chapter 4) were well ahead of their time and haven't received the attention they deserve.

Another issue is the relative value of theory vs. experiments. Since my own research has involved both theory and experiments, I don't feel I have a natural bias toward one or the other. It certainly is the case that evolutionary ecology as a field has plenty of theory, not all of it terribly helpful. In many fields, like genetics, critical experiments take on much more value than theory. For instance, we are all familiar with the classical experiment by Avery et al. (1944) which established that DNA not proteins or lipids was the hereditary material. However, there is little notice of the first person to suggest that DNA might be the hereditary material. I do consider critical experimental tests as extremely important, especially for a field in which difficult-to-test theories abound. However, this book has certainly recognized many of the important contributions of theory to the development of ideas and experiments in evolutionary ecology. I also include a number of laboratory experiments that I think have been especially useful in forming our understanding of evolutionary ecology principles. I have previously commented on the roles of experiments vs. studies of natural populations (Mueller and Joshi, 2000, Chapter 1) and won't review those issues here.

The area inviting the author's subjectivity is the inclusion of a paradigm-shift score, which was started in the first volume in this series (Avise, 2014). Here the goal is to assign some level of importance for each breakthrough. This includes many aspects of the works long-term influence, the breadth of the works impact and for theories the extent to which they have been supported by empirical work. By the time I was finished writing I realized that rather than use the entire impact scale of 1 through 10, as Avise (2014) and Breed (2017) had done, my scores were only in the range 5−10. I found it hard to give any work I felt deserved inclusion on this book the lowest scores. Alas, anyone unsatisfied with my apparent grade-inflation may convert one of my scores (m) to the Avise/Breed scale using the linear transformation, $round\{1+(9/5)(m-5)\}$, where "$round\{\}$" implies rounding off a fraction to the nearest integer.

Having pled guilty to allowing some subjectivity into my own analysis of the work included in this book, let me offer all the outstanding evolutionary ecologists whose work I have overlooked, my sincere apologies.

Finally, I would like to thank John Avise, John Thompson, Joe Travis, and David Reznick for helpful suggestions and discussions about their own views of what are the important contributions to evolutionary ecology.

REFERENCES

Avery, O.T., MacCeod, C.M., McCarty, M., 1944. Studies of the chemical nature of the substance inducing transformation of pneumococcal types. J. Expt. Med. 79, 137−158.

Avise, J.C., 2014. Conceptual Breakthroughs in Evolutionary Genetics. Academic Press, San Diego.

Breed, M.D., 2017. Conceptual Breakthroughs in Ethology and Animal Behavior. Academic Press, San Diego.

Mueller, L.D., Joshi, A., 2000. Stability in Model Populations. In: Monographs in Population Biology. Princeton University Press, Princeton, NJ.

CHAPTER ONE

1859 And in the beginning

The concept

There can be no serious discussion of evolutionary ecology without a reference to Charles Darwin and his revolutionary theory of evolution by natural selection. These ideas were originally laid out in his book "On the origin of species" (Darwin, 1859). Darwin insightfully recognized how the physical and biological environment will interact with variation within species to produce adaptations that permit organismal survival in those environments. Indeed, much of the focus of evolutionary ecology has been the detailing of how those interactions have proceeded and molded important physiological, morphological, and life historical traits of organisms.

The explanation

Darwin's writing was simple yet precise and his keen insights about the impact of the effects of local ecology can best be explained through his own words. His idea was that the environment was not constant and thus environmental variation was a challenge organisms must adapt to. Selection in variable environments has been a major topic of study by evolutionary ecologists.

> *...we do not always bear in mind, that though food may be now superabundant, it is not so at all seasons of each recurring year.*
>
> **Darwin (1859, chpt 3)**

Darwin also understood that a population could not increase exponentially without bound. He used this concept to suggest that the struggle for existence, and hence the opportunity for natural selection, could happen during any part of the organism's life cycle starting with the seed or egg.

> *Every being, which during its natural lifetime produces several eggs or seeds, must suffer destruction during some period of its life, and during some season or occasional year, otherwise, on the principle of geometrical increase, its numbers would quickly become so inordinately great that no country could support the product.*
>
> **Darwin (1859, chpt 3)**

Conceptual Breakthroughs in Evolutionary Ecology
ISBN: 978-0-12-816013-8
https://doi.org/10.1016/B978-0-12-816013-8.00001-6

Life-history evolution has become a prominent part of evolutionary ecology. Certainly, as I will discuss later, the timing of reproduction and mortality events has been a focus of both theoretical and experimental research in evolutionary ecology.

In a simple way Darwin also showed how quantitative and population dynamic reasoning could be brought to bear on questions of evolution.

> *Linnaeus has calculated that if an annual plant produced only two seeds—and there is no plant so unproductive as this—and their seedlings next year produced two, and so on, then in twenty years there would be a million plants.*
>
> **Darwin (1859, chpt 3)**

Of all the biological sciences, evolution and evolutionary ecology have a long history of utilization of mathematical theories. Darwin clearly understood this and used calculations of this sort to explicitly demonstrate the ease with which organisms could overpopulate their environment.

The physical environment can produce unique challenges to life. Darwin understood that organisms better able to tolerate the stresses of heat, cold, and humidity extremes would have an advantage.

> *The action of climate seems at first sight to be quite independent of the struggle for existence; but in so far as climate chiefly acts in reducing food, it brings on the most severe struggle between the individuals, whether of the same or of distinct species, which subsist on the same kind of food.*
>
> **Darwin (1859, chpt 3)**

Thus, competition may be intensified by abiotic factors.

> *...if these enemies or competitors be in the least degree favoured by any slight change of climate, they will increase in numbers, and, as each area is already fully stocked with inhabitants, the other species will decrease.*
>
> **Darwin (1859, chpt 3)**

Darwin also noted how the composition of a community has effects on the species present. So, when Scotch Fir was introduced to a stretch of barren heath plantations he noted,

> *...but twelve species of plants (not counting grasses and carices) flourished in the plantations, which could not be found on the heath. The effect on the insects must have been still greater, for six insectivorous birds were very common in the plantations, which were not to be seen on the heath; and the heath was frequented by two or three distinct insectivorous birds.*
>
> **Darwin (1859, chpt 3)**

Evolution of predators and prey has been another important area of research in evolutionary ecology as has the study of competing species.

A major problem which Darwin was unable to solve was the nature of the hereditary material. He proposed that this material was carried in particles called gemmules which circulated through the blood and collected in the reproductive organs. This idea was tested by Francis Galton (1871) by transfusing blood between different varieties of rabbits. Needless to say the experiment provided clear evidence against Darwin's theory of heredity.

Impact: 10

Darwin's work is not only the most important scientific breakthrough for evolutionary ecology but arguably for all of science. The fact that we can find in Darwin's *Origin of Species* descriptions of important problems that are still the focus of study today lends credence to the vast impact his work has had.

References

Darwin, C., 1859. On the Origin of Species by Means of Natural Selection, or the Preservation of Favored Races in the Struggle for Life. John Murray, London.

Galton, F., 1871. Experiments in pangenesis by breeding from rabbits of pure variety into whose circulation blood taken from other varieties had been infused. Proc. R. Soc. Lond. 19, 404.

CHAPTER TWO

1894 Measuring selection in nature

The concept

As the work of Darwin excited the community of scientists there was interest in collecting evidence of natural selection in action. Walter F. R. Weldon was one of the first to take on this task. Darwin emphasized the role of individual variation that might confer an advantage in reproduction or survival to its carrier. Weldon made measurements of the shell size of the crab *Carcinus moenas* among young and adult crabs (Weldon, 1894–1895). By separating out changes that might be due to growth versus changes that were due to selective loss he made one of the first attempts to measure natural selection in nature through differential survival.

The explanation

Motivated by Darwin, Weldon reasoned that the primary consequence of the struggle for existence would be reduced mortality among certain individuals that have traits that are now well adapted to their environment. Not knowing what these traits might be, Weldon decided to make extensive measurements of the shell morphology of young and adult crabs and see if these changed. His inference was that changes would be the evidence of selective deaths as predicted by Darwin. To carry out this work, Weldon needed to assume that as crabs grew the relative proportions of dimensions like carapace length and frontal breadth would change in concert. This may in fact not be correct and with standards in place today a preliminary experiment under protected conditions to show this was the case would be needed. Nevertheless, Weldon went on to suggest that for certain categories of crabs there was evidence of selective deaths. This work was the start of a long tradition for empirical tests of Darwin's theory of natural selection. The work is also notable for the detailed statistical analysis applied to the measurements collected by Weldon.

Conceptual Breakthroughs in Evolutionary Ecology
ISBN: 978-0-12-816013-8
https://doi.org/10.1016/B978-0-12-816013-8.00002-8

Impact: 5

By itself Weldon's work doesn't retain any lasting discovery. However, as an early precursor for the search for evidence of natural selection in nature it a good example of how this work could be done.

Reference

Weldon, W.F.R., 1894–1895. An attempt to measure the death-rate due to the selective destruction of *Carcinus moenas* with respect to a particular dimension. Proc. R. Soc. Lond. 57, 360–379.

CHAPTER THREE

1920 A theory of density-dependent population growth is formulated

The concept

Raymond Pearl and Lowell Reed (1920) derived a simple, two-parameter model to describe population growth. Their model is called the logistic equation and has been used extensively in ecological and evolutionary theory as a simple description of density-dependent population growth.

The explanation

Pearl and Reed (1920) approached the development of a simple model of population growth from a very practical perspective. They were interested in predicting the total size of the population of the United States at years between the census periods, which were every 10 years. Methods in use at the time assumed that the population would continue to grow at the same rate observed in the two most recent census periods in either an arithmetic or geometric fashion. Pearl and Reed realized that for long-term forecasting, these methods were inadequate. In fact, their best estimate of population growth in 1920 predicted a US population size of over 11 billion in the year 3000. Pearl and Reed concluded this population density would be absurd.

To arrive at a model of population growth that might make accurate predictions over long time periods, Pearl and Reed argued from first principles that such a model must have certain characteristics. The population size at time-t, N_t, should approach some positive asymptote as $t \rightarrow \infty$. By this they are requiring that the population not grow without bound but has some equilibrium. They also required the model have a second asymptote at 0 as $t \rightarrow -\infty$. Additionally, they required that the growth rate be proportional to the current population size and that the "still unutilized potentialities of population support". If the population uses all of its "potentialities" at

Conceptual Breakthroughs in Evolutionary Ecology
ISBN: 978-0-12-816013-8
https://doi.org/10.1016/B978-0-12-816013-8.00003-X

the positive equilibrium, then the last requirement suggests that growth will be proportional to how far below this equilibrium the current population size is.

Pearl and Reed (1920) arrive at the relationship, $N_t = \frac{be^{at}}{1+ce^{at}}$. This equation has become know as the logistic equation. If we let $a = r$ and $a/c = K$ then Pearl and Reed derive the derivative of N_t with respect to t $\left(\frac{dN_t}{dt}\right)$ as, $rN_t\left(\frac{K-N_t}{K}\right)$. This is the usual form the logistic is found in recent literature. The parameter r is the intrinsic rate of increase and represents the rate of population growth at low population density. K is the carrying capacity and represents the number of individuals at the high-density equilibrium. Estimating the parameters from US census data from 1790 to 1910 Pearl and Reed estimate the carrying capacity of the US population to be 197,274,000. In 2010 the US census reported a total population of just over 308 million. However, since 1910, advances in medicine, agriculture, and demographic patterns of childbirth would have certainly affected both r and K in the logistic equation rendering the Pearl and Reed estimates inaccurate.

Impact: 10

While overly simplistic, the logistic equation has served as a simple starting place for ecological and evolutionary modeling to the current day. The ubiquitous use of this relationship has given this work its very high impact.

Reference

Pearl, R., Reed, L.J., 1920. On the rate of growth of the population of the United States since 1790 and its mathematical representation. Proc. Natl. Acad. Sci. U.S.A. 6, 275–288.

CHAPTER FOUR

1928 Selection in age-structured populations

The concept

In 1918, R. A. Fisher settled an important debate in population genetics when he showed how biometrical measurements like correlations between relatives could be derived from standard Mendelian genetic models. However, Fisher's work treated populations with discrete generation life-histories. Future research on life-history evolution would require some understanding of how natural selection works in populations with age-structure. H. T. J. Norton's seminal paper in 1928 (Norton, 1928) developed many of the basic results for selection in age-structured populations. Serious work in this area would not begin again Charlesworth, 1970). Norton's most important result was to show that the outcome of selection at a single-locus could be predicted from the genotypic intrinsic rates of increase.

The explanation

Genetic variation for survival and fertility of genotype (A_iA_j) at a single locus can be described by the survival function $l_{ij}(x)$ and the fertility function $m_{ij}(x)$ which describe the chance of surviving to age-x and number of offspring produced at age-x, respectively. This formulation assumes no differences between the sexes and a maximum possible lifespan of d. Then the intrinsic rate of increase for A_iA_j is r_{ij} which is given as the real, positive root, z, of, $\int_0^d e^{-zx} l_{ij}(x) m_{ij}(x) dx = 1$. Norton showed that if A_1 carrying genotypes were inferior and possibly recessive $(r_{11} \le r_{12} < r_{22})$ or possibly dominant $(r_{11} < r_{12} \le r_{22})$ then selection would result in the frequency of A_1 going to 0. In the case of overdominance Norton concluded there might be a stable equilibrium or cycles. The possibility of cycles was ruled out by Charlesworth (1994).

Conceptual Breakthroughs in Evolutionary Ecology
ISBN: 978-0-12-816013-8
https://doi.org/10.1016/B978-0-12-816013-8.00004-1

Impact: 8

Norton was far ahead of his time but his early work solved some of the very difficult mathematical problems concerning age-structured population genetics. This work was eventually largely completed and tied into modern problems of life-history evolution by Charlesworth (1994). According to Haldane (1927), Norton started his work in this paper in 1910 and was mostly completed by 1922.

References

Charlesworth, B., 1970. Selection in populations with overlapping generations. I. The use of Malthusian parameters in population genetics. Theor. Popul. Biol. 1, 352–370.

Charlesworth, B., 1994. Evolution in Age-Structured Populations. Cambridge University Press, Cambridge.

Haldane, J.B.S., 1927. A mathematical theory of natural and artificial selection. Part IV. Math. Proc. Camb. Philos. Soc. 23, 607–615.

Norton, H.T.J., 1928. Natural selection and Mendelian variation. Proc. Lond. Math. Soc. 28, 1–45.

CHAPTER FIVE

1930 The fundamental theorem of natural selection

The concept

Sir Ronald Fisher (1930) showed that natural selection at a single locus will increase the mean fitness of the population and that increase will be proportional to the additive genetic variance for fitness. He called this result "The Fundamental Theorem of Natural Selection". This theorem has been invoked to suggest that natural selection will always result in a population being better adapted to its environment.

The explanation

Fisher is credited with many substantial accomplishments in population genetics and statistics. His 1918 paper (Fisher, 1918) is credited with showing how Mendelian genetic systems could explain phenotypic correlation between relatives. This work united the disparate groups of Mendelian and Biometrician branches of genetics (Provine, 1971). However, his book that was first published in 1930 (Fisher, 1930) sought to give a larger view of population genetics and evolution. Indeed, the latter third of the book is devoted to human genetics (which I will briefly touch on later).

One of Fisher's most important results was his demonstration that natural selection would increase the mean fitness of the population at a rate proportional to the additive genetic variance. Fisher's discussion and proof of this is somewhat opaque and I will outline here a much more direct proof from Crow (1986). The most detailed analysis of the properties of selection at a single locus were worked out by Kingman (1961).

Imagine a single locus with two alleles, A_1 and A_2, at frequencies p_1 and p_2 respectively. The fitnesses of the three genotypes, A_1A_1, A_1A_2, and A_2A_2 are w_{11}, w_{12}, and w_{22} respectively. Then the population mean fitness, $\overline{w}$, is, $p_1w_1 + p_2w_2$, where w_1 is $p_1w_{11} + p_2w_{12}$ and w_2 is $p_1w_{12} + p_2w_{22}$.

Conceptual Breakthroughs in Evolutionary Ecology
ISBN: 978-0-12-816013-8
https://doi.org/10.1016/B978-0-12-816013-8.00005-3

Then after one generation of selection the standardized change in mean fitness, $\frac{\Delta\overline{w}}{\overline{w}}$, is.

$$\frac{\Delta\overline{w}}{\overline{w}} \approx \frac{2\left[p_1(w_1-\overline{w})^2+p_2(w_2-\overline{w})^2\right]}{\overline{w}^2} \approx \frac{V(w)}{\overline{w}^2}$$

where $V(w)$ is the additive variance in fitness. The importance of this theorem was that it gave justification for evolutionary biologists to claim that evolution would provide maximal adaptation. I discuss the limitations of this claim in more detail in Chapter 26.

Fisher was a major participant in the early development of theoretical population genetics and evolution. He clearly had lofty expectations and thought that biology could use mathematics with the same skill as physicists. In the preface to his book Fisher (1930) longs for the time when "…there is built up a tradition of mathematical work devoted to biological problems, comparable to the researches upon which a mathematical physicist can draw in the resolution of special difficulties". While these hopes of Fisher have not been realized, the first seven chapters of "The genetical theory of natural selection" are all high-quality, important contributions to evolutionary theory.

In contrast, the last five chapters on human genetics and the decay of human civilization are today shocking in their racial, elitist tone. Fisher seemed to feel that social classes of people in modern society was a reflection of their genetic traits. Fisher was deeply concerned that higher status people in western civilization were not having as many children as people from the lower social classes. Fisher summarizes this point of view (Fisher, 1930, pg. 221–222) "… success in human endeavor is inseparable from the maintenance or attainment of social status; wherever, then, the socially lower occupations are more fertile, we must face the paradox that the biologically successful members of our society are to be found principally among its social failures, and equally that classes of persons who are prosperous and socially successful are, on the whole, the biological failures, the unfit of the struggle for existence, doomed more or less speedily, according to their social distinction, to be eradicated from the human stock." Fisher's inability to recognize the non-genetic, environmental contribution to "social status" is somewhat surprising given that Fisher was by all other measures a brilliant man.

Impact: 10

The fundamental theorem of natural section has had a lasting and important impact on research in evolutionary ecology and is thus given this high rating.

References

Crow, J.F., 1986. Basic Concepts in Population, Quantitative, and Evolutionary Genetics. W. H. Freeman, New York.

Fisher, R.A., 1918. The correlation between relatives on the supposition of Mendelian inheritance. Trans. R. Soc. Edinb. 52, 399–433.

Fisher, R.A., 1930. The Genetical Theory of Natural Selection. Oxford University Press.

Kingman, J.F.C., 1961. A mathematical problem in population genetics. Math. Proc. Camb. Philos. Soc. 57, 574–582.

Provine, W.B., 1971. The Origins of Theoretical Population Genetics. University of Chicago Press.

CHAPTER SIX

1930 Evolution of mimicry

The concept

Bates (1862) and Müller (1879) provided the first accounts of the evolutionary forces that might be responsible for the existence of species mimics. Fisher (1930) reviewed these and other explanations and proposed a sounder selection-based explanation for mimicry.

The explanation

Mimicry is the appearance of two different species that bear a striking similarity in appearance. This similarity can be striking. Bates (1862), when reviewing mimics of the *Heliconidae* butterflies remarks that "The resemblance is so close, that it is only after long practice that the true can be distinguished from the counterfeit, …". Bates concluded that the reason there are so many mimics of the *Heliconidae* is to benefit from the protection afforded the *Heliconidae* due to their bad taste. Mimics that lack such protection are now referred to as Batesian mimics; the species they resemble are referred as the model. Bates (1862) clearly postulated that these mimics have evolved these traits due to natural selection, "This principle can be no other than natural selection, the selecting agents being insectivorous animals …". Mimicry then is one of the earliest documented examples of evolution in response to the biotic environment attributed to natural selection after the appearance of Darwin's "Origin of Species"…

Müller (1879) on the other hand suggests that species that are both distasteful may also evolve to resemble each other since "… if two distasteful species are sufficiently alike to be mistaken for one another, the experience acquired at the expense of one of them will likewise benefit the other…". Today we recognize Batesian mimics as species with no special protection that bear a striking resemblance to a model which has some natural protection due to taste, stinging capability, or some other means of deterring predators. Müllerian mimics are species which resemble each other even though both possess a means of deterring predators.

Conceptual Breakthroughs in Evolutionary Ecology
ISBN: 978-0-12-816013-8
https://doi.org/10.1016/B978-0-12-816013-8.00006-5

Fisher (1930) provides an analysis of the evolutionary forces likely to be at work with these two types of mimicry. Fisher reasoned that Batesian mimicry will only persist if predators do not frequently encounter the mimic or the model is extremely noxious. If the mimic becomes common and predators begin feeding on models, there will be strong evolutionary pressure for the model to evolve different coloring. Fisher (1930) concludes that a stable Batesian mimic model system will typically require that the mimic be rare relative to the model to keep intact the reinforcement of predator avoidance.

Fisher also addressed the evolution of Müllerian mimicry. At that time, one theory suggested that the evolutionary route of Müllerian mimicry would be that the less common of two distasteful protected species, let's call it species *B*, would evolve toward the common species, (species *A*), but the reverse could not happen. Fisher argued that when mutant or recombinant genetic variants of species *A* are produced, they are just as likely to be more similar to species *B* as they are to be more dissimilar. However, owing to the protection afforded to the genetic variants that tend to resemble species *B*, they can increase in the population. Hence, evolution of the more common species toward the less common species is possible.

Impact: 8

Mimicry is a dramatic and widespread phenomena. This phenomenon served as one of the early demonstrations of the power of natural selection. The work of Fisher (1930) would be followed forty years later by additional theoretical work on the evolution of mimicry (reviewed in Chapter 39).

References

Bates, H.W., 1862. Contributions to the insect fauna of the Amazon valley. Trans. Linn. Soc. 23, 495–566.

Fisher, R.A., 1930. The Genetical Theory of Natural Selection. Oxford University Press.

Müller, F., 1879. *Ituna* and *Thuridia*: a remarkable case of mimicry in butterflies. Trans. Entomol. Soc. Lond. xx–xxix.

CHAPTER SEVEN

1930 Fluctuation in numbers and genotypes

The concept

Ford and Ford (1930) make an early and interesting suggestion that population cycles may be linked to the appearance of new genetic variants.

The explanation

Ford and Ford (1930) used museum samples and their own observations to document the history of a population of a butterfly, *Melitaea aurinia,* from 1881 to 1930. Their records suggest that the study population showed a gradual increase in numbers until it peaked in 1896. At this time the butterflies also experienced a high level (~95%) of parasitism, which probably contributed to the population's decline. From 1912 to about 1920 butterflies were nearly impossible to find. There was then a rapid increase from 1920 to about 1925 which was followed by a slow increase until the time of the Ford's paper, 1930.

Ford and Ford also made observations concerning the variety of color and pattern variation on the butterfly's wings. These observations must be tempered by the fact that (i) the sample sizes were small and non-random, and (ii) the variation was not quantified but described in a qualitative manner, e.g. "darker", "smaller", etc. Ford and Ford (1930) report similar levels of variation prior to 1896 and after 1925 but an increased frequency of variation during the rapid population increase.

Ford and Ford reasoned that in stable populations, mutations have come to an equilibrium between their addition by new mutations and their removal by natural selection. They then suggest that a rapid increase in population size is an indication of the relaxation of selection. With this relaxation there should be an increase in the frequency of new mutants. This analysis turns on the assumption that selection acts only through mortality, a view commonly called hard selection (see Wallace, 1975 for an exposition of these concepts). However, it is now widely accepted that much genetic variation may be under soft selection. Examples of soft

Conceptual Breakthroughs in Evolutionary Ecology
ISBN: 978-0-12-816013-8
https://doi.org/10.1016/B978-0-12-816013-8.00007-7

selection are frequency-dependent selection where fitness differences vanish at an equilibrium. Another example is mating preferences, which can be under strong selection but are not expected to affect population size. A more likely explanation for the appearance of novel phenotypes is the effect of small population size and inbreeding creating unusual genotypes.

While the exact mechanism that Ford and Ford proposed was naïve, the paper makes the interesting connection between population size and genetic variation. These ideas would be taken up later by Pimentel (1961; Chapter 17) and Carson (1975) in his founder-flush theory of speciation.

Impact: 6

Ford and Ford (1930) made one of the earliest connections between population dynamics and genetic variation. This idea would become an important focus of research in evolutionary ecology.

References

Carson, H.L., 1975. Genetics of speciation at the diploid level. Am. Nat. 109, 83–92.

Ford, H.D., Ford, E.B., 1930. Fluctuation in numbers, and its influence on variation, in *Melitaea aurinia*, Rott. (Lepidoptera). Trans. Entomol. Soc. London 78, 345–351.

Pimentel, D., 1961. Animal population regulation by the genetic feed-back mechanisms. Am. Nat. 95, 65–79.

Wallace, B., 1975. Hard and soft selection revisited. Evolution 29, 465–473.

CHAPTER EIGHT

1934 Competitive exclusion

The concept

Georgii F. Gause studied interactions between species using both models of inter-specific competition and experimental tests with *Paramecium* (Gause, 1934). He concluded that two competing species occupying the same niche will result in one or the other species going extinct. This prediction became known as Gause's competitive exclusion principle.

The explanation

Gause was motivated to formally study Darwin's ideas about the struggle for existence. Gause begins his book, "The struggle for existence" by reviewing work on competition in plants but concludes that "Botanists have endeavored to investigate the struggle for existence … but they are only beginning to analyze these phenomena" (Gause, 1934, pg 19). Concerning animals his conclusions were even more severe, "As concerns animals we have simply no exact data …" (Gause, 1934, pg 19).

Gause was clearly influenced by Pearl and Reed's (1920) work on density-dependent population growth (see Chapter 3). This theory provided a simple means of quantifying the struggle for existence as populations become more crowded relative to their resources. The most natural way to study evolution under density-dependence would have been to use the logistic equation with different genotypes. However, this treatment would have to wait another 40 years. Gause instead chose to study competition between two different species.

Letting the numbers of species-1 be N_1 and the number of species-2 be N_2 Gause studied the systems of differential equations,

$$\frac{dN_1}{dt} = b_1 N_1 \frac{K_1 - (N_1 + \alpha N_2)}{K_1} \tag{8.1a}$$

$$\frac{dN_2}{dt} = b_2 N_2 \frac{K_2 - (N_2 + \beta N_1)}{K_2} \tag{8.1b}$$

Conceptual Breakthroughs in Evolutionary Ecology
ISBN: 978-0-12-816013-8
https://doi.org/10.1016/B978-0-12-816013-8.00008-9

where K_i is the carrying capacity or equilibrium density of species-i, α is the competition coefficient that measures the impact of species-2 on the growth of species-1 and β is the competition coefficient for species-1.

Gause does not do a complete stability analysis of the system (8.1a and 8.1b) but notes that these equations will not permit an equilibrium between the competing species if they occupy the same "niche". By this he presumably means that $\alpha = \beta = 1$ although he does not specifically say this. Instead he claims that "...with the usual α and β there cannot simultaneously exist positive values for both $N_{1,\infty}$, and $N_{2,\infty}$. One of the species must eventually disappear." This prediction has become known as Gause's competitive exclusion principle.

Gause also attempts to test his theory by competing different species of *Paramecium*, *P. aurelia*, and *P. caudatum*. Although Gause eventually finds conditions under which the two species do not coexist (*P. aurelia* displacing *P. caudatum*) it is not without some difficulty. The details of the experimental environment affect these results and Gause needs to manipulate the conditions of the experiment to arrive at a set under which one species is eliminated. Gause's struggle with these experiments highlight the difficulty with his principle – defining the niche and determining if two species use exactly the same niche.

Impact: 9

Gause's work was one of the very earliest attempts to combine theory and experimental research in ecology and evolution. The focus on competition as an organizing force in community structure would continue among ecologist for many years. However, as ecologists made more substantial efforts to quantify a species niche it became clear that showing two species are using the environment in exactly the same way is nearly impossible.

References

Gause, G.F., 1934. The Struggle for Existence. Hafner, London.

Pearl, R., Reed, L.J., 1920. On the rate of growth of the population of the United States since 1790 and its mathematical representation. Proc. Natl. Acad. Sci. U.S.A. 6, 275–288.

CHAPTER NINE

1940 The common garden experiment

The concept

Clausen et al. (1940) developed a research protocol for sorting out phenotypic differences that were environmentally induced (phenotypic plasticity) versus genetic. This technique utilized raising genetically different individuals in a common garden for one generation.

The explanation

A variety of Western American plants have a wide distribution along an altitudinal gradient that runs from sea level on the coast of California to the mountains toward the East. These plant populations or ecotypes differ in several phenotypes but it was not clear to Clausen et al. if these differences were due to genetic differences or the pronounced environmental differences. To test the basis of these phenotypic differences, Clausen et al. (1940) took plants of the same species from different locations and grew them in a common garden. These gardens were located at Stanford (30 m elevation), Mather (1400 m), and Timberline (3050 m).

Clausen et al. measured traits like stem length and date of first flower. They reasoned that if ecotypes grown in a common garden showed phenotypic differences, then these would indicate genetic differences. Likewise, when clones from the same individual were raised in the different common gardens, any phenotypic differences would reflect environmental effects. The techniques that Clausen et al. applied so rigorously in this work would be emulated by many ecologists and evolutionary biologists. Such techniques continue to be used today. Even when working with animals, scientists today will refer to doing a common garden experiment.

Impact: 10

This research paradigm is still used today for studies of phenotypic plasticity and genetic differences that arise in laboratory evolution studies.

Conceptual Breakthroughs in Evolutionary Ecology
ISBN: 978-0-12-816013-8
https://doi.org/10.1016/B978-0-12-816013-8.00009-0

Because of the enduring value of the common garden experiment it has the highest impact score.

Reference

Clausen, J., Keck, D.D., Hiesey, W.M., 1940. Experimental studies on the nature of species I. In: Effect of Varied Environments on Western North American Plants. Carnegie Institute of Washington Pub. No. 520, Washington, DC.

CHAPTER TEN

1943 Seasonal changes of gene regions in natural populations

The concept

Populations of *Drosophila pseudoobscura* are polymorphic for inversions on their third chromosome. Theodosius Dobzhansky (1943) made estimates of inversion frequencies at roughly monthly intervals for three years. These inversions showed marked seasonal cycles. Dobzhansky suggested that different inversions may contain combinations of alleles at multiple loci that make their carriers more fit at different seasons.

The explanation

In the 1940's, evolutionary biologists had none of the molecular techniques in use today to quantify levels of population genetic variation. Instead, techniques based on morphological characters or in the case of species like *Drosophila pseudoobscura*, chromosomal inversions that are polymorphic were used. Dobzhansky (1943) measured five different third chromosome inversions (Standard, Arrowhead, Chiricahua, Tree Line and Santa Cruz) in three localities over 4 years. The three localities in California range in altitude from 800 feet above sea level to 4300 feet.

Frequencies for two of the inversions are shown in Fig. 10.1. There is a pronounced seasonal variation in these frequencies, which lead Dobzhansky to suggest that these were not random fluctuations but represented combinations of alleles in each inversion which improved fitness in different seasons: "The available data seem to fit best a fourth hypothesis, which assumes that the carriers of different gene arrangements in the third chromosome have different ecological optima" (Dobzhansky, 1943).

Individuals carrying only one copy of an inversion will not give rise to recombinant progeny since recombination will produce inviable chromosomes with duplications or deletions. Thus, Dobzhansky saw these inversions as potential genetic reservoirs for adaptive mutations. Dobzhansky suggested in his 1943 paper that "A mutant gene which is advantageous at a certain season arises in a chromosome with, for example, the Standard

Conceptual Breakthroughs in Evolutionary Ecology
ISBN: 978-0-12-816013-8
https://doi.org/10.1016/B978-0-12-816013-8.00010-7

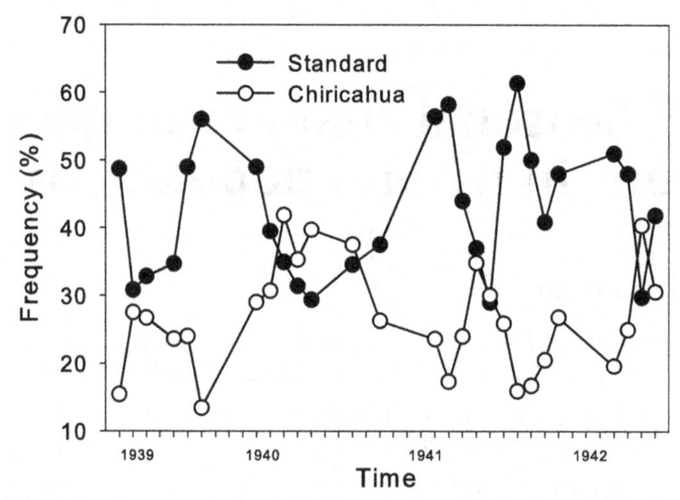

Fig. 10.1 Inversion frequencies from the Pinon, Mount San Jacinto, California site over 4 years (Dobzhansky, 1943).

arrangement. The descendants of this chromosome may attain a considerable frequency within the colony, and if the gene in question lies in the part of the chromosome in which crossing over is rare in inversion heterozygotes, the association may persist for a long time."

With time these inversions became known as co-adapted complexes of genes (Dobzhansky, 1950). An important test of inversions as co-adapted gene complexes was carried out by Strickberger (1963). He showed that inversions that had evolved in laboratory bottle environments apparently possessed favorable genetic variation that did not help them in cage environments relative to the same inversions that had been allowed to evolve in the cage environment.

Forty years later, single-gene polymorphisms in *D. pseudoobscura* and *Drosophila persimilis* would be followed for about 3.5 years (Mueller et al., 1985). These data did not show obvious seasonal cycles. However, Mueller et al. (1985) suggested that patterns of correlated change at pairs of unlinked loci could be used to determine if temporal variation in fitness played a role in the patterns of allele frequency variation over time. For a minority of loci pairs such patterns emerged.

Impact: 7

This work stimulated research in co-adapted gene complexes and more general work on the ability of selection across multiple loci to result in stable levels of linkage disequilibrium. The work was also important for showing how direct observation of populations in their natural environments might lead to clues about the mechanisms of natural selection.

References

Dobzhansky, T., 1943. Genetics of natural populations IX. Temporal changes in the composition of populations of *Drosophila pseudoobscura*. Genetics 28, 162–186.

Dobzhansky, T., 1950. Genetics of natural populations. XIX. Origin of heterosis through natural selection in populations of *Drosophila pseudoobscura*. Genetics 35, 288–302.

Mueller, L.D., Barr, L.G., Ayala, F.J., 1985. Natural selection versus random genetic drift: evidence from temporal variation in allele frequencies in nature. Genetics 111, 517–554.

Strickberger, M.W., 1963. Evolution of fitness in experimental populations of *Drosophila pseudoobscura*. Evolution 17, 40–55.

CHAPTER ELEVEN

1947 The role of phenotypic plasticity as an agent for adaptation to variable environments is proposed

The concept

Gause (1947) proposed that in a variable environment natural selection will not favor specialization to any particular environment but rather will favor increased phenotypic plasticity.

The explanation

The Russian ecologist G. F. Gause is best known for developing the competitive exclusion principle (Chapter 8, Gause, 1934). This principle asserts that no two species can exploit the environment in exactly the same way and coexist — one of the species will be excluded. However, in his 1947 paper, Gause explores in some detail the ability of animals to physiologically adjust to stressful environments.

Gause presents numerous experiments with *Paramecium*. He generated norms of reaction at different temperatures and salinities. Gause was struck by the fact that paleontologists suggest that many lineages of dinosaurs had evolved elaborate morphological defenses which ultimately made them less able to survive in a changing environment. Gause was informed by these observations and believed that organisms continue to be challenged this way. Is it better for a species to evolve genetic specializations to a narrow set of environmental conditions or should a species evolve a general flexibility in the form of phenotypic plasticity?

Gause (1947) has a separate section addressing "Adaptation to changing environments". In this section he concludes that "...natural selection under the conditions of a variable environment will favor the decrease of specialization at the expense of increased plasticity" (Gause, 1947).

Conceptual Breakthroughs in Evolutionary Ecology
ISBN: 978-0-12-816013-8
https://doi.org/10.1016/B978-0-12-816013-8.00011-9

Gause refers to numerous locations around Moscow where he obtained his samples of *Paramecium*. In postwar Soviet Union it is interesting to note that Gause repeatedly writes with great pride about the advances made by Russian scientists in his 1947 article.

Impact: 6

It would be some time before more rigorous studies of the role phenotypic plasticity as an adaptation to variable environments were pursued. Yet Gause's article is an early discussion of the important role of phenotypic plasticity in evolution.

References

Gause, G.F., 1934. The Struggle for Existence. Williams and Wilkins, Baltimore. Reprint, 1971, Dover, New York.

Gause, G.F., 1947. Problems of evolution. Trans. Conn. Acad. Arts Sci. 37, 17–68.

CHAPTER TWELVE

1947 Measuring selection and drift in a natural population

The concept

Using six consecutive years of allele frequency changes and population size estimates in the moth *Panaxia dominula*, Ronald A. Fisher and Edmund B. Ford (1947) tested the relative importance of selection and drift on allele frequency variation. While they found that the variation in allele frequencies was greater than expected from drift alone, their research marks the first of many efforts to sort out the evolutionary importance of drift and selection in natural populations.

The explanation

Sewell Wright (1931) emphasized the important role of random genetic drift in the evolutionary process. He suggested that in moderate-size populations drift would act to change allele frequencies in a direction that might not be favored by natural selection but could allow the population to explore the fitness landscape and possibly evolve to a higher fitness plateau. Fisher was not sympathetic to this view and the data collected by E. B. Ford allowed a direct comparison of the impact of selection and drift on allele frequency variation.

Ford had worked with a single locus wing color polymorphism in the moth *Panaxia dominula*. From 1941 to 1946, collections made at Oxford, UK provided yearly estimates of the frequency of the *medionigra* allele. The adult moths were mostly active during the month of July for 12–23 days. During this time, individual moths would be marked and then the frequency of recaptures would be recorded. These data allowed Fisher and Ford to estimate the size of the adult population.

Fisher realized that the estimated frequency of the *medionigra* allele would vary due to two different sampling processes; the finite breeding population (e.g. drift) and the number of adults sampled to estimate the *medionigra* allele frequency. Fisher then devised a statistical test to see if the allele frequency variation over a six year sampling period was about what was expected from

Conceptual Breakthroughs in Evolutionary Ecology
ISBN: 978-0-12-816013-8
https://doi.org/10.1016/B978-0-12-816013-8.00012-0

drift and allele frequency estimation or if it was greater than this, as might be expected if selection was acting on this polymorphisms but changing direction over time.

The estimates of population size were crude, so Fisher and Ford (1947) set the effective population size to be constant and at the lowest number seen over their six-year sample. They found that the observed variation was significantly greater than expected by drift and sampling alone. Fisher and Ford (1947) then concluded "Thus our analysis, the first in which the relative parts played by random survival and selection in a wild population can be tested, does not support the view that chance fluctuations in gene-ratios, such as may occur in very small isolated populations, can be of any significance in evolution." The last part of this statement is a little strong given that at best this study was on a single population. Nevertheless, this work showed the power of combining evolutionary theory, statistics, and detailed field observation in an evolutionary study.

Thirty years later, as many more datasets of allele frequency time-series in natural populations became available, Schaffer et al. (1977) published an improved version of Fisher and Ford's test. However, both Schaffer et al. (1977) and Fisher and Ford (1947) only had estimates of the census population size. Mueller et al. (1985) made actual estimates of effective population size from demographic and census data of butterfly populations and used computer simulations to derive drift and sampling-only expectations.

Impact: 8

This work is a pioneering application of modern statistical techniques to analyze the two major evolutionary forces of drift and natural selection in a natural population. The fact that a similar study did not happen for another 30 years shows how far ahead of its time Fisher and Ford's work was.

References

Fisher, R.A., Ford, E.B., 1947. The spread of a gene in natural conditions in a colony of the moth *Panaxia dominula* L. Heredity 1, 143–174.

Mueller, L.D., Wilcox, B.A., Ehrlich, P.R., Heckel, D.G., Murphy, D.D., 1985. A direct assessment of the role of genetic drift in determining allele frequency variation in populations of *Euphydryas editha*. Genetics 110, 495–511.

Schaffer, H.E., Yardley, D., Anderson, W.W., 1977. Drift or selection: a statistical test of gene frequency variation over generations. Genetics 87, 371–379.

Wright, S., 1931. Evolution in mendelian populations. Genetics 16, 97–159.

CHAPTER THIRTEEN

1954 Cole's paradox

The concept

A major division in life-history classifications are semelparous organisms which reproduce once in their lifetimes and then die versus iteroparous organisms which may reproduce multiple times after becoming sexually mature. In 1954 Lamont Cole (Cole, 1954) developed a simple model to compare population growth rates for a semelparous organism and an iteroparous organism. He concluded that a semelparous organism could equal the growth rate of an immortal iteroparous organism by simply producing one additional offspring. This result suggests that iteroparity should be a rare life-history strategy, yet in fact it is quite common and hence the paradox.

The explanation

Cole assumed that there was no juvenile mortality and hence the change in population size of a semelparous organism could be modeled as $N_{t+1} = F_s N_t$, where F_s is the number of offspring produced on average by each semelparous individual and N_t is the population size at time-t. Thus, the per-capita rate of increase in a single generation would be $\frac{N_{t+1}}{N_t} = F_s$. Cole then compared this to the growth rate of an iteroparous organism that starts reproduction at the same time as the semelparous organism, suffers no mortality as an adult, and produces F_I offspring at each time interval during its adult life-span. With those assumptions, the change in population size during each time interval would be, $N_{t+1} = F_I N_t + N_t$. The per-capita rate of growth is then, $F_I + 1$.

These results imply that the per-capita rates of growth will be equal if $F_s = F_I + 1$. This is a rather modest requirement. This surprising result stimulated a general inquiry into the conditions under which iteroparity would evolve. One set of assumptions that appeared severe was the lack of mortality among the juveniles and adults. Charnov and Schaffer (1973) suggested that these assumptions be relaxed. Thus, if juvenile survival in the semelparous and iteroparous organism was set to P_J and the survival of adults between reproductive bouts was set to P_A, then after some algebra it is found that

Conceptual Breakthroughs in Evolutionary Ecology
ISBN: 978-0-12-816013-8
https://doi.org/10.1016/B978-0-12-816013-8.00013-2

the semelparous per-capita growth rate will equal the iteroparous growth rate if, $F_s = F_I + \frac{P_A}{P_J}$. This relationship shows that semelparity will be favored if juvenile survival is high relative to adult survival. The reverse conditions will tend to favor iteroparity. We consider some additional explanations in the following chapters.

Impact: 6

While the results of Cole's calculations have not stood up to time, they did stimulate research to provide a quantitative understanding for those conditions that would favor iteroparity.

References

Charnov, E.L., Schaffer, W.M., 1973. Life-history consequences of natural selection: Cole's result revisited. Am. Nat. 107, 791–793.
Cole, L.C., 1954. The population consequences of life history phenomena. Q. Rev. Biol. 29, 103–137.

CHAPTER FOURTEEN

1954 Lack's principle

The concept

The number of eggs laid by birds does not seems to be at their physiological limit. David Lack (1954) proposed that birds limit the number of eggs they produce to the number they can successfully raise. Since newly hatched birds require parental care to successfully reach reproductive age, natural selection will favor maximizing the number of birds fledged rather than the number of eggs.

The explanation

In Lack's important book on the regulations of animal population size (Lack, 1954) he addresses the perplexing problem for ornithologists of what determines clutch size. Lack's arguments throughout his book are supported by an expansive reference to natural historical data. He first considers two non-evolutionary explanations: (1) *The number of eggs laid is at the physiological limit.* This is easily rejected since if eggs are removed from a nest birds typically lay additional eggs to replace them. (2) *The number is limited by how many a sitting bird can cover.* However, hatching percentage doesn't seem to vary over a large range of egg number. The third hypothesis is a group selection argument. (3) *Individual clutch size is adjusted to match the current population size and thus prevent over production of young*. However, birds don't appear to match their clutch size to current population densities.

A final argument by Lack was based on evolutionary principles. Since newly hatched birds require parental care to reach sexual maturity, fertility is a combination of both egg numbers and successful rearing of the young. Lack realized that if birds laid too many eggs, the number of young they could successfully raise may in fact decline. Thus, Lack suggested that clutch size would be tuned by natural selection to be as large as each species could be reasonably expected to raise to independence.

Lack also reflected on animals that can vary the number and size of their eggs. He suggests that natural selection may cause animals to lay fewer, larger eggs under conditions where food resources are scarce. Larger eggs should

Conceptual Breakthroughs in Evolutionary Ecology
ISBN: 978-0-12-816013-8
https://doi.org/10.1016/B978-0-12-816013-8.00014-4

provide larger offspring which will be able to survive better in resource poor areas.

Lack's focus in his book (Lack, 1954) is almost entirely on natural populations. However, while discussing density-dependence he refers to laboratory studies of Pearl (Pearl and Reed, 1920), Park (1941), and Nicholson (1950). Lack has a decidedly negative view of this experimental research which is reflected in his conclusion that he "...doubted whether the extensive research on laboratory populations is of much help in interpreting how animal numbers are regulated in nature,..." (Lack, 1954, pg. 18). This point of view is encountered frequently even among current research scientists. However, it misses the point that laboratory research is not intended to determine how populations of *Drosophila*, flour beetles, or blowflies are regulated in nature. Rather, they are used to test general theories of evolution or population dynamics. These issues are discussed in more detail in the first chapter of Mueller and Joshi (2000).

Impact: 7

Lack was an early adopter of evolutionary theory in his exploration of some traditional ecological problems. His approach was widely adopted and influential in the development of evolutionary ecology.

References

Lack, D., 1954. The Natural Regulation of Animal Numbers. Oxford.

Mueller, L.D., Joshi, A., 2000. Stability in Model Populations. Monographs in Population Biology. Princeton University Press, Princeton, NJ.

Nicholson, A.J., 1950. Population oscillation caused by competition for food. Nature 165, 476–477.

Park, T., 1941. The laboratory population as a test of a comprehensive ecological system. Q. Rev. Biol. 16, 274–293, 440–461.

Pearl, R., Reed, L.J., 1920. On the rate of growth of the population of the United States since 1790 and its mathematical representation. Proc. Natl. Acad. Sci. U.S.A. 6, 275–288.

CHAPTER FIFTEEN

1956 Character displacement

The concept

Species that are very similar in areas of allopatry may show divergence for several phenotypes in regions of sympatry. This divergence is driven by species competition.

The explanation

For some time evolutionary biologists have noted a repeated pattern among similar species. In areas of allopatry these species pairs will be remarkably similar but when they are sympatric they show a divergence in characters. These characters may be morphological, physiological, ecological, or behavioral. Brown and Wilson (1956) named these patterns "character displacement."

Brown and Wilson (1956) review a number of examples of possible character displacement in birds, ants, fish and frogs. They suggest that the similarity of the species in the areas of allopatry is due to their relatively recent speciation. The areas of sympatry are special since the similarity of the two species almost guarantees their competition for essential resources. The negative fitness consequences of this competition can be reduced if one or both species diverge in characters related to the resources for which they compete.

Although there are many patterns consistent character displacement, showing that this displacement was due to competition is much more challenging. We examine such a case in Chapter 59.

Impact: 7

Brown and Wilson (1956) marked the organized presentation of the verbal theory of character displacement and was the start of many important studies on this problem.

Reference

Brown Jr., W.L., Wilson, E.O., 1956. Character displacement. Syst. Zool. 5, 49–64.

Conceptual Breakthroughs in Evolutionary Ecology
ISBN: 978-0-12-816013-8
https://doi.org/10.1016/B978-0-12-816013-8.00015-6

CHAPTER SIXTEEN

1958 Niche partitioning by warbler birds

The concept

MacArthur (1958) undertook a detailed study of the ecology of sympatric species of warbler birds. What made this research stand out is MacArthur's ability to organize his observations around theories of competition and competitive exclusion.

The explanation

MacArthur studied five species of congeneric warblers that seemed so similar that "…ecologists studying them have concluded that any differences in the species' requirements must be quite obscure…" (MacArthur, 1958). However, MacArthur was guided by theories of competitive exclusion which dictated that if two or more species use the environment in the same way only one species will persist: "…to permit coexistence it seems necessary that each species, when very abundant, should inhibit its own further increase more than it inhibits the other's" (MacArthur, 1958).

MacArthur relies on published research and some indirect methods to first conclude that the five warbler populations are limited by density-dependent events and that in the case of the Cape May warbler the limiting resource was food. With this conclusion, MacArthur reasoned that any overlap in the use of limiting resources will lead to competition. To determine if such overlap occurs, MacAthur spent two summers (in 1956 and 1957) making detailed observations of the birds. To quantify his observations, he divided the volume of the spruce and fir trees where foraging and nesting took place into 16 intra-tree regions (such as trunk, inner branches, outer branches, etc.). These regions were divided by height above the ground and distance from the trunk. He then recorded how long each of the five species of warblers spent in specific foraging locations.

These records showed that each species had a distinct part of the tree where most of its foraging efforts were focused. It also turned out that many of the nesting sites for each bird were close to their preferred foraging

Conceptual Breakthroughs in Evolutionary Ecology
ISBN: 978-0-12-816013-8
https://doi.org/10.1016/B978-0-12-816013-8.00016-8

area. MacArthur also kept track of how long the birds flew between bouts of sitting and foraging. While in the trees some birds would hop from branch to branch while others would walk along a branch. MacArthur devised a method for recording the direction of movements while each bird was foraging so he could later compare these. MacArthur also observed differences in the duration of these flights and the types of movements made by birds while foraging.

MacArthur concluded that "First, the observations show that there is every reason to believe that the birds behave in such a way as to be exposed to different kinds of food. They feed in different positions, indulge in hawking and hovering to different extents, move in different directions through the trees, vary from active to sluggish, and probably have the greatest need for food at different times corresponding to the different nesting dates" (MacArthur, 1958). While any two species showed some degree of overlap for some of these aspects of feeding, altogether MacArthur concluded that the warbler behaviors allowed them to avoid direct intense competition and ultimately coexist. MacArthur did not explicitly develop an evolutionary explanation for these observations. This research was certainly a starting point for later research on the evolution of resource partitioning (see Chapter 40).

Impact: 9

MacArthur's work was an important collection of natural history guided by the impact of interspecific competition and the notion that species exclusion would be the outcome of severe competition. More formal theoretical research on partitioning resources to reduce levels of competition would be motivated by this work.

Reference

MacArthur, R.H., 1958. Population ecology of some warblers of Northeastern coniferous forests. Ecology 39, 599–619.

CHAPTER SEVENTEEN

1961 Population regulation and genetic feedbacks

The concept

David Pimentel suggested that in addition to density-dependence and random environmental fluctuations, animal populations may be controlled by a genetic feedback mechanism with their food source.

The explanation

To develop his basic idea of genetic feedbacks, Pimentel presents some examples where he believed such interactions may have taken place. One example is the Hessian fly, which is an introduced species that feeds on wheat. The Hessian fly rose to large numbers in wheat regions of the United States, such as Kansas, prior to 1942. Around that time, resistant genotypes of wheat were introduced and the numbers of Hessian flies dropped dramatically. Another example is a disease which struck the Oyster population of Malpeque Bay, Canada in 1915. The catch of Oysters dropped dramatically and remained low until about 1930 when it started a gradual rise due to resistance to the unidentified parasite. This resistance was not found in other Oyster populations.

While Ford and Ford (1930) had made similar suggestions linking population cycles to genetic changes within populations (Chapter 7), Pimentel (1961) took the suggestion further by developing a genetic and population dynamic model. Pimentel's model involved a plant-herbivore system. He modeled a single-locus genetic system in the plant population but did not model the plant population dynamics, in essence assuming there were an excess number of plants at all times. However, the allele frequencies at a single locus in the plant population were subject to change. The fitness of each plant genotype was determined as a fixed effect of genotype, independent of the animal population. There was a second effect on the fitness of each plant genotype which was a linear, decreasing function of the number of animals in the population. Plant genotypes varied in their sensitivity to herbivory. The animal population grew according to which

Conceptual Breakthroughs in Evolutionary Ecology
ISBN: 978-0-12-816013-8
https://doi.org/10.1016/B978-0-12-816013-8.00017-X

genotype of plant they consumed, with some plant genotypes leading to animal population growth, others leading to decline. There was no density-dependent regulation of the animal population. Finally, the animal population also had a single-locus, two-allele polymorphism, and the fitness of each of these three genotypes was a function of which genotype of plant they consumed.

While this model is moderately complicated, Pimentel grossly under-analyzed it. In fact, he only presented a single numerical example which resulted in an oscillation of the animal population size. There were no general results indicating the conditions for such oscillations or if they were in fact stable. Nevertheless, the basic idea was made – the genetics and numbers of individuals in different trophic levels may influence each other.

A more recent review of this issue was undertaken by Schoener (2011). A more detailed example of this interaction is reviewed in Chapter 54.

Impact: 5

Pimentel's theoretical work left a lot to be desired. However, by framing the idea of genetic feedback between interacting species, he opened up discussion of an exciting area of research in evolutionary ecology.

References

Ford, H.D., Ford, E.B., 1930. Fluctuation in numbers, and its influence on variation, in Melitaea aurinia, Rott. (Lepidoptera). Trans. Entomol. Soc. London 78, 345–351.

Pimentel, D., 1961. Animal population regulation by the genetic feed-back mechanisms. Am. Nat. 95, 65–79.

Schoener, T.W., 2011. The newest synthesis: understanding the interplay of evolutionary and ecological dynamics. Science 331, 426–429.

CHAPTER EIGHTEEN

1962 Ecological measures of fitness

The concept

By 1962 the development of population genetic theory was well advanced. Fitness in these models was assumed constant and acted as a measure of relative survival of a genotype from zygote to sexually mature adult. MacArthur (1962) proposed measuring the fitness of genotypes in the neighborhood of a population-size equilibrium or carrying capacity. With this theory MacArthur derived the equivalent of Fishers Fundamental Theorem of Natural Selection which suggests natural selection will increase a population's carrying capacity.

The explanation

The traditional theories of natural selection worked out by Haldane, Fisher, and Wright typically treated fitness as a constant. However, one aspect of the environment that ecologists knew was regularly changing was the population size. In fact, much of ecological theory had been devoted to modeling population growth. A model that received much attention was due to Pearl and Reed (1920) and became known as the logistic equation (Chapter 3).

The logistic equation contained two parameters, r, the intrinsic rate of increase and K, the carrying capacity. The carrying capacity is also an equilibrium point of the logistic model and is interpreted, ecologically, as the maximum density of individuals that can be sustained by the environment. MacArthur's contribution was to assert that fitness may in fact vary with population density and thus population dynamics may directly impact changes in allele frequencies. In addition, his work shows that adaptation by natural selection will alter the balance of births and deaths in resource-limited environments.

MacArthur accomplishes this by assuming that at a single locus with two alleles, each genotype's fitness (e.g. births minus deaths), is a function of the frequency of each allele and the total population size. MacArthur refers to

Conceptual Breakthroughs in Evolutionary Ecology
ISBN: 978-0-12-816013-8
https://doi.org/10.1016/B978-0-12-816013-8.00018-1

the logistic equation but does not formally use it in his analysis. Some of MacArthur's arguments get a little confused (for instance, when he asserts that the frequency of an allele is equal to the number of copies of the allele divided by the total population size rather than twice the population size). But his general conclusion can still be made that natural selection will favor the genotype able to have a positive growth rate at the highest density. MacArthur didn't state that evolution would increase the carrying capacity but rather "natural selection seems always to decrease the density of limiting resource required to maintain the population at a constant level" (MacArthur, 1962).

This paper initiated several decades of theoretical and empirical research into the evolution of population dynamics and life histories. Extensions of this theory were carried out by MacArthur and Wilson (1967), Anderson (1971), Roughgarden (1971), and Clarke (1972). These theories laid out in a more explicit fashion the genetic system and where in the life cycle density affects fitness. The consequences of complex life cycles for the predictions from this theory were explored by Prout (1980).

Impact: 10

The work of MacArthur in many ways marks the beginning of the modern era of evolutionary ecology. Making the fitness of a genotype sensitive to levels of population crowding was just one way in which ecological factors can be incorporated in evolutionary models. Many more would follow MacArthur's pioneering paper.

References

Anderson, W.W., 1971. Genetic equilibrium and population growth under density-regulation selection. Am. Nat. 105, 489–498.

Clarke, B., 1972. Density-dependent selection. Am. Nat. 106, 1–13.

MacArthur, R.H., 1962. Some generalized theorems of natural selection. Proc. Natl. Acad. Sci. U.S.A. 48, 1893–1897.

MacArthur, R.H., Wilson, E.O., 1967. Island Biogeography. Princeton University Press, Princeton.

Pearl, R., Reed, L.J., 1920. On the growth rate of the population of the United States since 1790 and its mathematical representation. Proc. Natl. Acad. Sci. U.S.A. 6, 275–288.

Prout, T., 1980. Some relationships between density-independent selection and density-dependent population growth. Evol. Biol. 13, 1–68.

Roughgarden, J., 1971. Density-dependent natural selection. Ecology 52, 453–468.

CHAPTER NINETEEN

1962 Group selection

The concept

Vero Wynne-Edwards proposed a novel mechanism for the evolution of altruistic traits. He suggested that traits that would not be favored by individual selection, but which may be beneficial to the entire group could nevertheless increase in frequency by the differential extinction of whole groups.

The explanation

Wynne-Edwards (1962, 1963) was concerned with population regulation and the mechanisms animals use to prevent overexploitation of their resources. He suggested that many social behaviors of animals were primarily adaptations to prevent overexploitation of resources. Among these behaviors are holding territories and social status. But Wynne-Edwards argued that these traits would be opposed by individual selection. After all individual selection ought to favor the consumption of as much of the available resources as possible and the production of as many offspring as possible. How then would any trait that is not favored by individual selection become established in a population? Wynne-Edwards argued that populations without traits that prevent overexploitation would go extinct due these behaviors leaving behind populations with the favorable traits. This process has become known as group selection.

Wynne-Edwards (1963) would also assert that group selection might also be responsible for the sociality seen in the insect groups of the Hymenoptera (but see Chapter 21). Wynne-Edwards also suggested his theory could be thought of as an extension of Wright's shifting balance theory. However, Wright's theory did not involve the differential extinction of whole populations and has at its core individual selection.

Using an experimental system of flour beetles, Michael Wade (1977) showed that if individual selection and group selection acted in the

Conceptual Breakthroughs in Evolutionary Ecology
ISBN: 978-0-12-816013-8
https://doi.org/10.1016/B978-0-12-816013-8.00019-3

opposite directions then sufficient levels of population extinction could overcome individual selection. However, most of the debate over group selection has revolved around how often the conditions allow for sustained differential extinction of whole populations. In addition, the development of kin selection helped explain the evolution of many altruistic behaviors (Chapter 21).

Impact: 5

Wynne-Edward's ideas were thought provoking and stimulated research on the evolution of social behavior and altruism via individual selection.

References

Wade, M.J., 1977. An experimental study of group selection. Evolution 31, 134–153.
Wynne-Edwards, V.C., 1962. Animal Dispersion in Relation to Social Behavior. Hafner, New York.
Wynne-Edwards, V.C., 1963. Intergroup selection in the evolution of social systems. Nature 200, 623–626.

CHAPTER TWENTY

1964 Community evolution

The concept

In 1964, Ehrlich and Raven (1964) noted that there was little research into the evolution of interacting species within a community that would not ordinarily exchange genetic information. They proposed that one way to study such evolution would be to focus on "... the examination of patterns of interaction between two major groups of organisms with a close and evident ecological relationship, such as plants and herbivore" (Ehrlich and Raven, 1964). They called this type of evolution, "coevolution". Since 1964 the study of coevolution has exploded, stimulated in large part by Ehrich and Raven's seminal paper.

The explanation

An important insight for Ehrlich and Raven was that plants often produce an array of chemicals that do not appear to serve any important physiological function but repel many insects and other herbivores. Alternatively, they noted that some insects have the ability to feed on plants that are almost universally rejected by other herbivores.

For a phytophagous animal there can be nothing more important than getting adequate food to develop and ultimately reproduce. Likewise, plants may suffer large reductions in fitness, even death, by high levels of herbivory. The opportunity for evolution to act on both plants and herbivores seemed a constant reality. Much of Ehrlich and Raven's article is devoted to the exhaustive review of the superfamily of butterflies (Papilionoidea) and their host plants. Ehrlich and Raven outline one route by which these taxa may evolve in response to each other. That is, they develop a verbal theory of coevolution of plants and their herbivores.

Ehrlich and Raven note that if a mutation arises in a plant to produce a novel, toxic compound, then the plant will enter a new adaptive zone. The new genotype will be able to live and reproduce in areas that were previously inaccessible due to herbivory. These plants will also present an opportunity for herbivores as a food source which is unexploited by other

Conceptual Breakthroughs in Evolutionary Ecology
ISBN: 978-0-12-816013-8
https://doi.org/10.1016/B978-0-12-816013-8.00020-X

herbivores. Thus, a mutant herbivore that can develop a tolerance for this toxic compound will itself have a new adaptive zone. In addition, this new herbivore may itself become unpalatable due to its ingestion of these toxins.

Impact: 10

Extending the view of evolution to interacting species has had substantial impact on evolutionary research. Since Ehrlich and Raven, coevolution has been applied to other species that show antagonistic interactions (such as parasites and their hosts), and also to mutualistic interactions, like plants and their pollinators (Thompson, 1994). It is hard to overstate the importance of this seminal paper.

References

Ehrlich, P.R., Raven, P.H., 1964. Butterflies and plants: a study in coevolution. Evolution 18, 586–608.

Thompson, J.N., 1994. The Coevolutionary Process. University of Chicago Press, Chicago.

CHAPTER TWENTY ONE

1964 Kin selection and the evolution of social behavior

The concept

Many animals exhibit altruistic behavior characterized by improvements in the fitness of another individual at the expense of the individual performing the altruistic act. Such behavior seems contrary to the normal rules of natural selection. Wynne-Edwards (1963) had suggested that such behavior may evolve if it improves the chances of group survival even if individual fitness is lowered. William Hamilton (1964a,b) developed the concept of inclusive fitness and showed that under certain circumstances rearing the young of or helping a close relative could be favored by evolution even if it bore a cost to the altruist. This theory was based on the usual rules of selection at the level of the individual and did not require the differential survival of whole groups.

The explanation

Hamilton (1964a,b) was interested in exploring whether an organism should engage in activities that benefit relatives other than their own offspring. His reasoning was that these relatives carry common alleles that they have both inherited from their common ancestor. The only difference

Table 21.1 Coefficient of relatedness for diploid and haplo-diploid individuals.

Genetic relatedness involving	Related pair	Diploid	Haplodiploid
Males	Father-son	0.5	0
	Mother-son	0.5	0.5
	Brothers	0.5	0.5
	Brother-sister	0.5	0.25
Females	Mother-daughter	0.5	0.5
	Full-sib sisters	0.5	0.75
	Half-sib sisters	0.25	0.25

Conceptual Breakthroughs in Evolutionary Ecology
ISBN: 978-0-12-816013-8
https://doi.org/10.1016/B978-0-12-816013-8.00021-1

between relatives such as siblings, cousins etc. is the degree of sharing. If an individual, called an altruist, orients a behavior or activity towards a relative, called the recipient, then there will be fitness gains and costs. Suppose the fitness gain to the relative equals B but entails a cost to the altruist equal to C. Under what conditions would this behavior evolve? Hamilton (1964) showed that if the coefficient of relatedness (the probability that an allele in the altruist is identical by descent to the allele in the recipient) is greater than C/B, then the behavior will be favored by natural selection. This was an extremely powerful result. For instance, it helps to explain why social insects like the Hymenoptera have sterile female workers that forgo reproduction to help raise their sisters. It turns out that Hymenoptera have a haplo-diploid genetic system in which females are diploid but males are haploid (Table 21.1). Thus, sisters inherit exactly the same set of genes from their father and they have a 50% chance of inheriting the same genes from their mother. Consequently, haplo-diploid sisters are more closely related to each other than are diploid parents and offspring (Table 21.1).

Kin selection, unlike group selection, can account for the concentration of social behavior in the Hymenoptera. Since kin selection works through the mechanisms of individual selection, it does not depend on the more problematic mechanism of differential extinction of whole populations.

Impact: 10

The impact of Hamilton's kin selection theory was substantial. It brought clarity to the Darwinian incongruity of altruism, and it may be one of the most consequential breakthroughs in evolutionary theory in the 20th century.

References

Hamilton, W.D., 1964a. The genetical evolution of social behavior. I. J. Theor. Biol. 7, 1–16.

Hamilton, W.D., 1964b. The genetical evolution of social behavior. II. J. Theor. Biol. 7, 7–52.

Wynne-Edwards, V.C., 1963. Intergroup selection in the evolution of social systems. Nature 200, 623–626.

CHAPTER TWENTY TWO

1965 Evolution of phenotypic plasticity in plants

The concept

Plants share the inability to move when their local environment changes in a substantial way. Bradshaw (1965) spelled out the many circumstances under which phenotypic plasticity might evolve in plant populations as a way of coping with environmental fluctuations.

The explanation

The idea that phenotypic plasticity could be molded by natural selection had been made earlier than Bradshaw's 1965 review article by Gause (1947) and Schmalhausen (1949). However, Bradshaw's paper brought special attention to plants. He notes that unlike animals plants do not have the option of moving if the environmental conditions change.

Bradshaw reviews several conditions under which phenotypic plasticity might be expected to evolve. Bradshaw notes that Clausen et al. (1940) have shown that there is variation in phenotypic plasticity response between populations, suggesting some tuning by natural selection. Perhaps the most important type of selection would be what Bradshaw called disruptive selection. By this Bradshaw suggests that over time the conditions in the same location may vary and thus favor different phenotypes. Alternatively, conditions may vary over space and thus favor different phenotypes within the same population.

Bradshaw also suggests that even if selection is directional for some phenotype, phenotypic plasticity could be advantageous. That is, after genetic variation has pushed the phenotype as far as possible, further gains might be made from a phenotypically plastic response. One of Bradshaw's less convincing arguments is that multiple genotypes may show the same phenotype due to plasticity of the trait. He suggests that this may be a way of maintaining genetic variation in populations but provides no quantitative support for this claim. Nevertheless, Bradshaw produces strong

Conceptual Breakthroughs in Evolutionary Ecology
ISBN: 978-0-12-816013-8
https://doi.org/10.1016/B978-0-12-816013-8.00022-3

arguments that many phenotypically plastic traits in plants may have been fine tuned by selection.

Impact: 6

This paper was important for keeping a focus on the adaptive basis of phenotypic plasticity and providing a detailed descriptive basis for how such selection might work. This work would be taken in much more quantitative direction later by Via and Lande (1985).

References

Bradshaw, A.D., 1965. Evolutionary significance of phenotypic plasticity in plants. Adv. Genet. 13, 115–155.

Clausen, J., Keck, D.D., Hiesey, W.M., 1940. Experimental studies on the nature of species I. Effect of varied environments on Western North American plants. Carnegie Institute of Washington Pub. No. 520. Washington, D. C.

Gause, G.F., 1947. Problems of evolution. Trans. Conn. Acad. Arts Sci. 37, 17–68.

Schmalhausen, I.I., 1949. Factors of Evolution: The Theory of Stabilizing Selection. Blakiston, Philadelphia, PA.

Via, S., Lande, R., 1985. Genotype-environment interaction and the evolution of phenotypic plasticity. Evolution 39, 505–522.

CHAPTER TWENTY THREE

1965 Fitness estimation

The concept

Evolutionary theory was subjected to empirical research in the middle of the 20th century. Estimating model parameters (such as fitness) turned out to be more complicated than most scientists expected. Prout (1965) showed that except for the simplest organisms, evolutionary models need to take into account an organism's life cycle and the various ways fitness can affect survival and reproduction at each stage of the life cycle.

The explanation

In the first half of the 20th century, great strides were made by scientists like Fisher, Wright, and Haldane developing the theory of population genetics and evolution. Fitness parameters in models of natural selection were treated as abstract constants that reflected the relative success of a genotype to survive and reproduce. However, it wasn't until the second half of the 20th century that attempts were made to estimate fitness empirically.

Except for unicellular organisms, most plants and animals have life cycles which present evolution with several opportunities to generate fitness differences. For instance, fruit flies can be made to reproduce on a simple discrete-generation cycle. Eggs are produced, juveniles develop to sexual maturity, males and females mate, and then eggs are collected over one short period, (e.g. one day), to start the next generation. Natural selection might affect survival from egg to adult and it may also affect the number of offspring that males and females produce due to differential fertility or mating success. Prout (1965) showed that these life-cycle complexities must be taken into account.

Prout (1965) showed that if the population allele frequencies are observed over one generation at a life stage before selection has completed, then fitness estimates will be erroneous. For example, suppose allele frequencies among adults at a single locus with two alleles are followed for one generation. Suppose that the relative egg-to-adult survival of the three genotypes (A_1A_1, A_1A_2, and A_2A_2) are 0.2, 1, and 1.5, respectively, and

Conceptual Breakthroughs in Evolutionary Ecology
ISBN: 978-0-12-816013-8
https://doi.org/10.1016/B978-0-12-816013-8.00023-5

fertilities of the three genotypes are 1.5, 1, and 0.2. Then the net fitness of the three genotypes, A_1A_1, A_1A_2, and A_2A_2, is the product of the early (survival) and late (fertility) fitness components, or 0.3, 1, and 0.3, respectively. However, experimentally we might follow the change in the A_1 frequency over one generation among adults before fertility selection has acted. Prout (1965) showed that if the adults start with a frequency of A_1 at 0.8, then the standard methods of estimating fitness would erroneously yield 0.31, 1, and 0.96 for A_1A_1, A_1A_2, and A_2A_2 respectively. Prout (1971a,b) lays out a protocol for properly estimating fitness which separately measures different components of fitness (such as viability, female fertility, and male mating success). Prout (1980) also extended this work to models of population growth and selection (see appendix).

Impact: 9

Prout's research developed an important link between theoretical evolution and empirical research. While theoreticians prefer to ignore the idiosyncrasies of organisms, Prout emphasized that for empiricists, life-history details are important.

References

Prout, T., 1965. The estimation of fitness from genotypic frequencies. Evolution 19, 546–551.

Prout, T., 1971a. The relation between fitness components and population prediction in Drosophila. I. The estimation of fitness components. Genetics 68, 127–149.

Prout, T., 1971b. The relation between fitness components and population prediction in Drosophila. II. Population prediction. Genetics 68, 151–167.

Prout, T., 1980. Some relationships between density-independent selection and density-dependent population growth. Evol. Biol. 13, 1–68.

CHAPTER TWENTY FOUR

1966 The concept of energetic trade-offs

The concept

In 1966, Martin Cody developed a theory to explain the variation in clutch size among birds. He describes the organizing principle by suggesting "It is possible to think of organisms as having a certain limited amount of time or energy available for expenditure, and of natural selection as that force which operates in the allocation of this time or energy in a way which maximizes the contribution of a genotype to following generations." This idea has been adopted and used to understand many problems in evolutionary ecology besides clutch size.

The explanation

Why can't organisms live forever, start reproducing soon after their birth or live on virtually any food resource? Certainly, these adaptations would make any organism better adapted to its environment. Cody's idea was that natural selection can't accomplish everything due to constraints. Cody chose to focus on an organism's total energy budget as providing constraints on a number of competing functions. Cody focused on the energy required by birds for reproduction, competition, and predator avoidance. Under this theory, as the requirement of one function (say competition) decreases, then energy can be reallocated to the other functions (like reproduction). Thus, if there is less competition among birds for resources in temperate regions, then birds can increase their energy expenditure on reproduction and hence increase their clutch size.

Evolutionary ecologists soon realized that the idea of competing energy demands, and their resulting trade-offs could be used to understand a number of phenomena. For instance, Gadgil and Bossert extended Cody's ideas to life-history evolution by noting "We may then set up the problem of life-history strategy as that of optimal allocation of resources among maintenance, growth and reproduction, just as Cody (1966) did as among clutch

Conceptual Breakthroughs in Evolutionary Ecology
ISBN: 978-0-12-816013-8
https://doi.org/10.1016/B978-0-12-816013-8.00024-7

size, predator avoidance, and competitive ability" (Gadgil and Bossert, 1970). The idea of trade-offs has been applied to a number of topics including evolution at extreme population densities (Mueller and Ayala, 1981), dispersal (Zera and Denno, 1997), reproduction and survival (Bell and Koufopanou, 1986), fecundity and survival (Callow, 1973) and many more.

Impact: 9

This early, explicit discussion of trade-offs in evolutionary ecology has generated a theme that has been repeated many, times. Often Cody does not receive credit for this idea but his works stands as a seminal contribution to the field.

References

Bell, G., Koufopanou, V., 1986. The cost of reproduction. Oxf. Surv. Evol. Biol. 3, 83–131.

Calow, P., 1973. The relationship between fecundity, phenology, and longevity: a systems approach. Am. Nat. 107, 559–574.

Cody, M., 1966. A general theory of clutch size. Evolution 20, 174–184.

Gadgil, M., Bossert, W.H., 1970. Life historical consequences of natural selection. Am. Nat. 104, 1–24.

Mueller, L.D., Ayala, F.J., 1981. Trade-off between *r*-selection and *K*-selection in *Drosophila* populations. Proc. Natl. Acad. Sci. U.S.A. 78, 1303–1305.

Zera, A.J., Denno, R.F., 1997. Physiology and ecology of dispersal polymorphism in insects. Annu. Rev. Entomol. 42, 207–230.

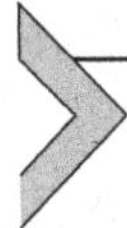

CHAPTER TWENTY FIVE

1966 Evolution of demographic parameters

The concept

William D. Hamilton (1966) sought to develop a simple understanding of how natural selection acts on equivalent changes in mortality as a function of age. He came up with the simple result that starting at sexual maturity, the strength of selection acting on equivalent changes in mortality will decrease with increasing age. Thus, senescence is a logical byproduct of natural selection.

The explanation

Shortly after his work on kin selection, Hamilton (1966) explored the action of natural selection in age-structured populations. He adopted the Malthusian parameter (m), utilized by both by Norton (1928) and Fisher (1930), as the measure of fitness in an age-structured population, which is the positive root, m, of the Lotka-Euler equation, $\int_0^\infty e^{-mx} l_x f_x = 1$, where l_x is the probability of surviving to age-x and f_x is the fecundity of females aged-x. If age is broken into discrete classes, 1, 2, ..., d, and p_x is the chance of an individual surviving from age-class x to age-class $x+1$, $l_x = p_1 p_2 \ldots p_{x-1}$. Hamilton looked at changes in p_x caused by mutations that increase or decrease survival by a factor $1-\delta$. From this type of analysis he found that similar changes in age-specific survival at different ages would always be most strongly favored (in the case of an increase in survival) at earlier ages. He concluded from this model that "A basis for the theory that senescence is an inevitable outcome of evolution is thus established". A test of this theory is described in Chapter 50.

Baudisch (2005) has recently suggested that there are other ways to compare mutations at different ages which won't necessarily give the same results. For instance, age-specific survival may be increased by an exponential factor, e.g. $p_x^{(1+\delta)}$. In this case, natural selection might resist senescent decline if fecundity changes with age in an appropriate direction. There

Conceptual Breakthroughs in Evolutionary Ecology
ISBN: 978-0-12-816013-8
https://doi.org/10.1016/B978-0-12-816013-8.00025-9

are at present few well documented cases of organisms which do not show increasing mortality with age. Some studies of natural populations proport to show this (Jones et al., 2014), but such studies in nature encounter severe technical issues (see appendix).

The evolutionary principles set forth by Hamilton have more recently been used to study late-life mortality plateaus. Carey et al. (1992) and Curtsinger et al. (1992) studied very large cohorts of laboratory fly populations and showed that although these populations displayed the expected increase in mortality through much of their adult life, at extremely old ages these mortality rates leveled off. These plateaus were an unexpected anomaly. However, Mueller and Rose (1996) demonstrated that such plateaus are an expected byproduct of the evolutionary forces identified by Hamilton plus random genetic drift. The impact of Hamilton's (1966) work has been reviewed by Rose et al. (2007).

Impact: 10

The theory developed by Hamilton has had lasting impact and has been empirically supported in numerous experiments (Rose et al., 2007). As such it is one of the foundational breakthroughs in evolutionary biology.

References

Baudisch, A., 2005. Hamilton's indicators of the force of selection. Proc. Natl. Acad. Sci. U.S.A. 102, 8263–8268.

Carey, J.R., Liedo, P., Orozco, D., Vaupel, J.W., 1992. Slowing of mortality rates at older ages in large medfly cohorts. Science 258, 457–461.

Curtsinger, J.W., Fukui, H.H., Townsend, D.R., Vaupel, J.W., 1992. Demography of genotypes: failure of the limited life span paradigm in *Drosophila melanogaster*. Science 258, 461–463.

Hamilton, W.D., 1966. The moulding of senescence by natural selection. J. Theor. Biol. 12, 12–45.

Jones, O.R., Scheuerlein, A., Salguero-Gomez, R., Giovanni Camarda, C., Schaible, R., Casper, B.B., Dahlgren, J.P., Ehrlen, J., Garcıa, M.B., Menges, E.S., Quintana-Ascencio, P.F., Caswell, H., Baudisch, A., Vaupel, J.W., 2014. Diversity of ageing across the tree of life. Nature 505, 169–173.

Mueller, L.D., Rose, M.R., 1996. Evolutionary theory predicts late-life mortality plateaus. Proc. Natl. Acad. Sci. U.S.A. 93, 15249–15253.

Rose, M.R., Rauser, C.L., Benford, G., Matos, M., Mueller, L.D., 2007. Hamilton's forces of natural selection after forty years. Evolution 61, 1265–1276.

CHAPTER TWENTY SIX

1966 Optimal foraging based on time and energy

The concept

If simple assumptions are made about the time a forager spends traveling, searching for food and the relative intake of energy, then predictions can be made about the breadth of a forager's diet, and when it should move to a new patch to seek food. These predictions assume that natural selection will favor those genotypes that optimize energy intake and that behaviors will reflect this optimization.

The explanation

There was a pervasive idea in evolutionary ecology that Fisher's Fundamental Theorem of Natural Selection (Chapter 5) could be used as a guide to the outcome of the evolutionary process. Thus, if natural selection carries a population to a local maximum in mean fitness, then a trait like foraging behavior should evolve to maximize fitness by maximizing energy intake. This assumption is not often stated explicitly. For instance, in their study of the optimal use of patches, MacArthur and Pianka (1966) state that "Hopefully, natural selection will often have achieved such optimal allocation of time and energy expenditures, but such 'optimum theories' are hypotheses for testing rather than anything certain". There is less tentativeness in Charnov (1976) when he simply says, "The predator is assumed to make decisions so as to maximize the net rate of energy intake during a foraging bout". I will return to these issues later.

MacArthur and Pianka (1966) explored the conditions under which a forager would add additional items to its diet or additional patches to its regular search routine. Their method for exploring these problems was simple, "The basic procedure for determining optimal utilization of time or energy budgets is very simple: an activity should be enlarged as long as the resulting gain in time spent per unit food exceeds the loss" (MacArthur and Pianka, 1966). To determine the optimal diet breadth MacArthur and Pianka compared the time spent searching (which should decline as you add

Conceptual Breakthroughs in Evolutionary Ecology
ISBN: 978-0-12-816013-8
https://doi.org/10.1016/B978-0-12-816013-8.00026-0

more items to your diet) to the time spent pursuing the item. This last factor assumes we are dealing with a predator and it assumes that more difficult-to-catch prey are added to the diet last. In the case of adding more patches, MacArthur and Pianka compare the change in hunting time per food item (by adding an additional patch to their search routine) to the change in search time per item by adding this additional patch.

Charnov (1976) focused on the problem of when a forager decides to leave the patch in which it is currently feeding. Charnov assumed that the amount of energy a forager can extract from a patch will level off as it depletes food. But leaving a patch entails an energy cost (so should not be done hastily). Charnov comes up with the simple rule that "The predator should leave the patch it is presently in when the marginal capture rate in the patch ($\partial g/\partial T$) drops to the average capture rate for the habitat".

Evolutionary biologists have struggled with the utility and role of optimization models (see Chapter 8 in Oster and Wilson, 1978 for a good discussion). Even if we allow that foraging efficiency is equivalent to fitness, we know that fitness is not always maximized. Kojima and Kelleher (1961) and Moran (1964) showed that population mean fitness may actually decline in two-locus selection models. This result was generalized by Karlin (1975) who showed that any two-locus model with a stable polymorphic equilibrium and linkage disequilibrium will not be at a local fitness maximum.

Putting the issue of fitness maximization aside there is still the issue of assuming that natural selection could optimize foraging behavior independent of other life history traits. Indeed, some experiments have shown this is not the case. Fruit flies kept under crowded conditions, where food is limiting, evolve to become less efficient at turning food into biomass rather than becoming more efficient as might be expected under an optimization view of evolution (Mueller, 1990; Joshi and Mueller, 1996). In the case of fruit flies, adaptation to crowding also involves frequency-dependent selection for increased competitive ability which is negatively correlated with efficiency of food use.

Oster and Wilson (1978) suggest that optimization theory is no more than a tactical tool for making educated guesses about evolution. However, they envision the experimental evolutionary biologist as the key in testing and rejecting theories with the classic strong-inference paradigm (Platt, 1964). But the role of the experimental evolutionary biologist is even more challenging since it is likely that many optimization models with the proper "tuning" can make essentially identical predictions. Designing critical experiments to sort through these is then the real challenge.

Impact: 9

MacArthur, Pianka, and Charnov started a new way of investigating the behavior of animals which continues to this day.

References

Charnov, E.L., 1976. Optimal foraging: the marginal value theorem. Theor. Popul. Biol. 9, 129–136.

Joshi, A., Mueller, L.D., 1996. Density-dependent natural selection in *Drosophila*: trade-offs between larval food acquisition and utilization. Evol. Ecol. 10, 463–474.

Karlin, S., 1975. General two-locus selection models: some objectives, results and interpretations. Theor. Popul. Biol. 7, 364–398.

Kojima, K., Kelleher, T.M., 1961. Changes of mean fitness in random mating populatinos when epistasis and linkage are present. Genetics 46, 527–540.

MacArthur, R.H., Pianka, E.R., 1966. On optimal use of a patchy environment. Am. Nat. 100, 603–609.

Mueller, L.D., 1990. Density-dependent natural selection does not increase efficiency. Evol. Ecol. 4, 290–297.

Moran, P.A.P., 1964. On the nonexistence of adaptive topographies. Ann. Hum. Genet. 21, 383–393.

Oster, G.W., Wilson, E.O., 1978. Social insects. In: Monographs in Population Biology, vol. 12. Princeton Univ. Press, Princeton, N. J.

Platt, J.R., 1964. Strong inference. Science 146, 347–353.

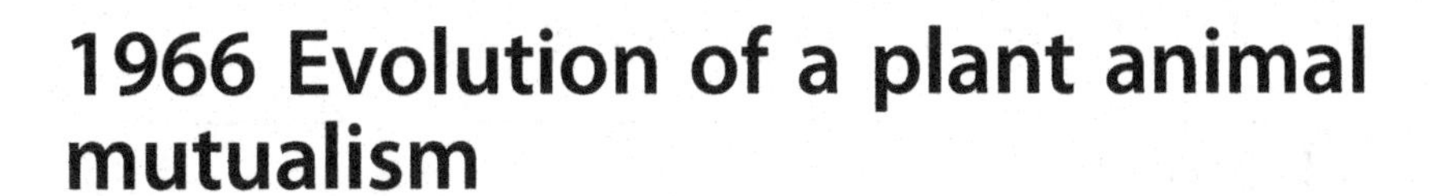

1966 Evolution of a plant animal mutualism

The concept

Interactions between species that benefit the interacting species are called mutualisms. Daniel Janzen (1966) provided one of the earliest and most extensive examples of a mutualism between the plant *Acacia cornigera* and the ant *Pseudomyrmex ferruginea.*

The explanation

Acacia cornigera which is also referred to as a swollen-thorn acacia is found in close association with ants (*Pseudomyrmex ferruginea*). Janzen (1966) undertook an extensive study of the interaction between these two species in their natural habitats of Central America. The acacia provides a home and food resources for the ants. The ants provide the acacias with protection from both insect herbivores and plant competitors.

The ants hallow out the large thorns of the acacia and use them as protective structures and a place for raising their young. The acacia also provides the ants food from their enlarged foliar nectaries. Acacias also produce small protein rich buds at the tip of their leaves called Beltian bodies which are an additional source of food for the ants. Virtually all the food used by the ant population comes from the combination of acacia nectar and Beltian bodies. Many acacias will shed their leaves during the dry seasons to conserve moisture. The swollen-thorn acacias provide nearly year-round leaf production to accommodate the ant population.

The ants are aggressive and will attack other insects and small mammals that attempt to forage on the acacia. Likewise, neighboring plants that start to encroach on the acacia or the space close (within 10–150 cm) to the acacia are trimmed down by the ants. This reduces the competition for growth and allows the acacias to increase in size more rapidly, which then also provides the ants with greater resources.

Janzen (1966) notes that one of the advantages of this system for studying a mutualism is that the ants can be experimentally removed from the acacias

Conceptual Breakthroughs in Evolutionary Ecology
ISBN: 978-0-12-816013-8
https://doi.org/10.1016/B978-0-12-816013-8.00027-2

to observe the impact that a lack of ants has on plant growth and survival. Janzen in fact predicts that the acacia population would probably go extinct if the local ant population were completely removed.

Impact: 10

Janzen's work remains as one of the best and most thoroughly documented examples of mutualism.

Reference

Janzen, D.H., 1966. Coevolution of mutualism between ants and acacias in Central America. Evolution 20, 249–275.

CHAPTER TWENTY EIGHT

1967 Evolution following colonization

The concept

In MacArthur and Wilson's (1967) book on island biogeography, they lay out the evolutionary changes expected to follow colonization. In this chapter they lay out the early foundations of the verbal theory of *r*- and *K*-selection. This theory dominated the discussion of life history evolution for many years and is still referred to in introductory textbooks.

The explanation

MacArthur and Wilson start their discussion by reviewing conclusions made earlier by MacArthur (1962) that in an expanding, uncrowded environment *r* from the logistic equation would reflect fitness, and in a crowded environment *K* will be the appropriate measure of fitness. Following this, MacArthur and Wilson expand upon MacArthur's earlier work by giving a verbal description of the phenotypes most likely to increase either *r* or *K*. In an uncrowded environment harvesting the most food (even if wastefully) should be favored to allow the largest families to be produced. Also, *r*-selected species should have a shorter development time, longer reproductive life and higher fecundity. Evolution under crowded conditions looks quite different. Here efficiency is at a premium. Selection will favor genotypes which can at least replace themselves on the lowest level of food, hence efficiency of converting food to offspring is at a premium.

In this same chapter MacArthur and Wilson (1967) also explore the conditions under which species might go through character displacement to reduce competition for a common resource. MacArthur and Wilson's work was followed by a number of studies that developed the conditions that would allow coexistence of competing species (MacArthur, 1970) and how resource utilization might evolve (Roughgarden, 1972). This leads to another important aspect of MacArthur and Wilson's book. In addition to these topics the book also, for the first time, develops a quantitative theory to describe the ecology of islands. This theory continues to be important today

Conceptual Breakthroughs in Evolutionary Ecology
ISBN: 978-0-12-816013-8
https://doi.org/10.1016/B978-0-12-816013-8.00028-4

in conservation biology. The breadth of interesting ideas in ecology and evolution explored by MacArthur and Wilson was exceptional making, this book a landmark in the field of evolutionary ecology.

Impact: 10

In terms of stimulating new research this book is perhaps one of the most influential in the field of evolutionary ecology.

References

MacArthur, R.H., 1962. Some generalized theorems of natural selection. Proc. Natl. Acad. Sci. U.S.A. 48, 1893–1897.

MacArthur, R.H., 1970. Species packing and competitive equilibrium for many species. Theor. Popul. Biol. 1, 1–11.

MacArthur, R.H., Wilson, E.O., 1967. Island Biogeography. Princeton Univ. Press, Princeton.

Roughgarden, J., 1972. Evolution of niche width. Am. Nat. 106, 683–718.

CHAPTER TWENTY NINE

1968 Cole's paradox resolved by environmental variation

The concept

Cole's paradox (Chapter 13) had been festering for nearly 18 years when Garth Murphy (Murphy, 1968) developed a simple genetic model in which a semelparous genotype competed with an iteroparous genotype. He found that if juvenile survival was highly variable, the iteroparous genotype would increase and eliminate the semelparous genotype. In a particularly bad year there would be few surviving progeny and thus very few semelparous individuals would make it to the next round of reproduction. However, iteroparous adults might survive these bad years and have another opportunity to reproduce. Iteroparous individuals could be viewed as hedging their bets and in a variable environment this turns out to be an advantageous reproductive strategy.

The explanation

Murphy had originally written his paper to focus on competition between different species with different life-histories. However, at the insistence of Monte Lloyd, he included a genetic model. This was important since it addressed more directly the possibility of evolution of iteroparity in a diploid, sexually reproducing organism. The genetic model was verbally described but not formally written out. Below I develop formal equations for Murphy's model.

Murphy assumed a single locus with two alleles, *A* and *a*. *AA* and *Aa* genotypes lived only one time interval and then died after reproduction. The *aa* homozygote lived for two time intervals and reproduced in each time interval. To keep track of the iteroparous individuals let $N_{aa}^{(1)}(t)$ and $N_{aa}^{(2)}(t)$ be iteroparous individuals alive at time-*t*, aged 1, and 2, time units respectively and $N_{aa}(t)$ be the sum of the two. Likewise, let $N_{Aa}(t)$ be the number of heterozygous, and $N_{AA}(t)$ the number of homozygous, semelparous individuals at time-*t*. Murphy set a limit of 10,000 individuals in the environment. Thus, the total offspring from all three genotypes that were

Conceptual Breakthroughs in Evolutionary Ecology
ISBN: 978-0-12-816013-8
https://doi.org/10.1016/B978-0-12-816013-8.00029-6

allowed to survive in any time unit was 10,000 minus the number of 1-unit old iteroparous individuals or, $T(t) = 10,000 - N_{aa}^{(1)}(t)$. The relative fecundity of the three genotypes, AA and Aa and aa was assumed to be $f/2$, $f/3$, and $f/6$ respectively where $f = 3T(t)/p_A(t)$ and $p_A(t)$ is the current frequency of the A-allele. With this notation then we can derive the following population size and allele frequency recursions,

$$N_{aa}^{(1)}(t+1) = (1 - p_A(t+1))^2 T(t), \tag{1a}$$

$$N_{aa}^{(2)}(t+1) = N_{aa}^{(1)}(t), \tag{1b}$$

$$N_{Aa}(t+1) = 2p_A(t+1)(1 - p_A(t+1))T(t) \tag{1c}$$

$$N_{AA}(t+1) = p_A^2(t+1)T(t) \tag{1d}$$

$$p_A(t+1) = \frac{\frac{1}{2}N_{AA}(t)f + \frac{1}{6}N_{Aa}(t)f}{\frac{1}{2}N_{AA}(t)f + \frac{1}{3}N_{Aa}(t)f + \frac{1}{6}N_{aa}(t)f}. \tag{1e}$$

Murphy (1968) uses computer simulations to see the outcome of this evolution. In a constant environment the semelparous allele, A, is fixed in 32 time units.

To simulate a random environment Murphy (1968) made set $T(t) = T(t)$ δ, where δ was a random variable that took on values of 1 or 0.1, presumably each with probability 0.5 although Murphy doesn't specify this. Murphy apparently only ran the random environmental simulation twice. In the first run the iteroparous a allele fixed in 12 time units and in the second run it took 25 time units. 21st century academic standards would have required these simulations be run about 1000 times or more so the entire distribution of results could be observed. However, in 1968 the use of computer simulations in evolutionary theory was still relatively new. I have run simulations using the equations above but also allowing the lower limit of δ to range from 0.1 to 0.001 (Fig. 29.1). As the environment gets more severe, e.g. δ gets smaller, the iteroparity allele is fixed more frequently (Fig. 29.1). Unlike Murphy's results I saw no cases out of 1000 trials that fixed the iteroparity allele when the lower limit on δ was 0.1. This is possibly due to assumptions I made (like how frequently the environment would be good or bad) that Murphy did not specify in his paper.

Murphy also varied the way fecundity was regulated. However, the basic results remained unchanged. Constant environments favored semelparity while variable environments generally favored iteroparity. Cohen (1966,

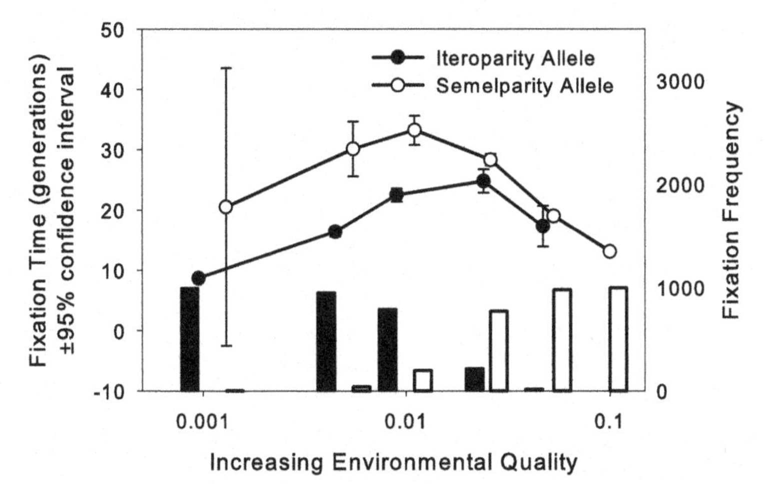

Fig. 29.1 The results of 1000 independent trials of evolution in a randomly varying environment using Eq. 1a–e. The right *y*-axis shows the number of times out of 1000 that the iteroparity allele (black fill) and the semelparity allele (white fill) were fixed as a function of environmental quality, e.g. the lower limit to δ. The left *y*-axis gives the average number of time units required for fixation. Each trial was run for 100 times units and occasionally neither allele was fixed. The probability of $\delta = 1$ was 50% each time unit.

1967) had earlier considered a related problem. He examined the effects of environmental variation in the production of seeds. Cohen found that in a randomly varying environment that affected seed production, plants would leave more offspring by making some of their seeds dormant, so they could germinate in the future when perhaps conditions might be better.

Impact: 8

As a theoretical paper the work of Murphy (1968) is not terribly elegant. But his idea brought great attention to the impact of environmental variation on life history evolution. Much of this work was improved by future studies like Schaffer (1974).

References

Cohen, D., 1966. Optimizing reproduction in a randomly varying environment. J. Theor. Biol. 12, 119–129.

Cohen, D., 1967. Optimizing reproduction in a randomly varying environment when a correlation may exist between the conditions at the time a choice has to be made and the subsequent outcome. J. Theor. Biol. 16, 1–14.

Murphy, G.I., 1968. Pattern in life history and the environment. Am. Nat. 102, 391–403.

Schaffer, W.M., 1974. Optimal reproductive effort in fluctuating environments. Am. Nat. 108, 783–790.

1968 Evolution in changing environments

The concept

Richard Levins (1968) developed a very general theory to assess evolution in the face of environmental variation. His development of the fitness set analysis was novel and allowed for solutions to complicated problems not previously addressed by evolutionary biologists.

The explanation

Bacteria can use the enzyme β-galactosidase to capitalize on galactose as a source of energy. However, it may take the bacteria 30 min to induce appreciable levels of the enzyme. Is the bacterium better off always making the enzyme or only turning it on when needed? Clearly, a bacterium that always makes the enzyme gets a head start of 30 min over any competitors that induce the enzyme when galactose first appears. However, if galactose is infrequently encountered, constitutive production of the enzyme will be costly and reduce fitness.

This is the sort of problem Levins addressed. Levins conjectured that for significant environmental changes the phenotypes best adapted in one environment might do poorly in another. If one could measure fitness across all possible phenotypes in each of two environments, the results might look something like Fig. 30.1 (left panel). Now if the fitness of each phenotype in environment-1 and environment-2 are used to create a set of x-y coordinates, the coordinates can be plotted in two dimensions (Fig. 30.1, right panel). This collection of fitness coordinates was designated the "fitness set". For the example in Fig. 30.1 the fitness set is convex. If the optimal phenotypes (e.g. the one with the highest fitness) are further apart, the fitness set is concave (Fig. 30.2).

A second important component of this theory was the frequency with which the organisms encountered each environment. Here we can think of the environment varying over time or space. If the organism spends most or all of its lifetime in one environment, then the environment is referred

Conceptual Breakthroughs in Evolutionary Ecology
ISBN: 978-0-12-816013-8
https://doi.org/10.1016/B978-0-12-816013-8.00030-2

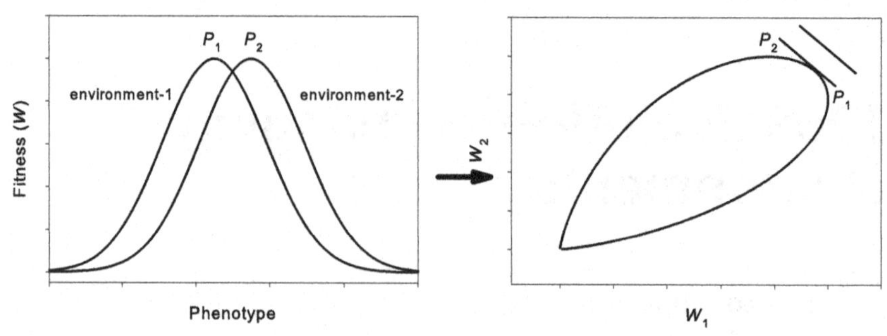

Fig. 30.1 The fitness in two environments over a continuous set of phenotypes. The phenotype with the highest fitness in environment-1, P_1, is different than the most fit phenotype in environment-2, P_2. Each phenotype on the x-axis can be assigned a fitness in environment-1 and a fitness in environment-2 and plotted as in the right figure. The oblong figure at the right is the fitness set. In this case the fitness set is convex (any line drawn between two points on the fitness set will be wholly contained in the fitness set). The fitnesses corresponding to P_1 and P_2 are shown on the fitness set. Adaptive functions for a fine-grained environment are shown as straight lines in the right figure. The point at which these lines touch the fitness set indicate the optimal phenotype which in this case is somewhere between the optimum in environment-1 and environment-2.

to as coarse-grained. If each environment may be encountered frequently during the organism's lifetime, then the environment is fine-grained.

Given details of the granularity of the environment Levins (1968), showed that an adaptive function could be described that quantifies the fitness values of populations with mixtures of each phenotype. These adaptive functions will appear as lines (Fig. 30.1, right panel) in the fitness set diagrams for fine-grained environments and curved lines for coarse-grained environments (Fig. 30.2, right panel).

When the fitness set is convex, the optimal outcome of evolution is a single phenotype which lies between P_1 and P_2. However, with a concave fitness set the outcomes are varied. In a fine-grained environment, it may be best to adapt via either P_1 or P_2. Alternatively, in a coarse-grained environment, the best solution is a polymorphic population with both the P_1 and P_2 phenotypes (Fig. 30.2, right panel).

Impact: 8

In practice this is a difficult theory to test. Nevertheless, Levins offered a unique perspective on the analysis of evolution in the face of environmental uncertainty.

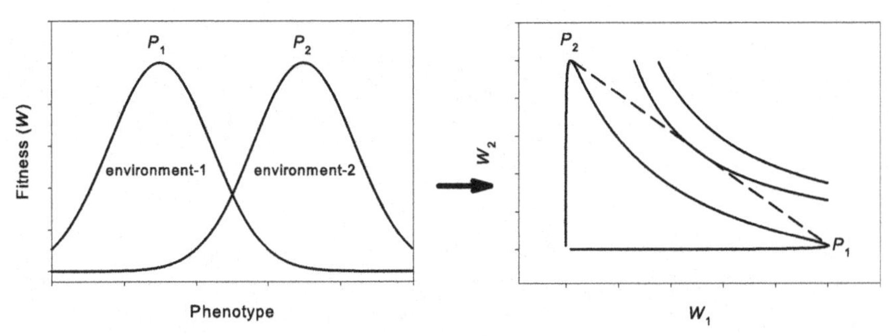

Fig. 30.2 The left figure is the same as in Fig. 30.1 except now the phenotypes with the highest fitness in the two environments are further apart. The result of this is a concave fitness function shown on the right. The dashed line joining the points P_1 and P_2 represent polymorphic solutions in which different mixtures of P_1 and P_2 are present in the population. The curved lines represent a coarse-grained adaptive function and where is touches the dashed line is the evolutionary optimum.

Reference

Levins, R., 1968. Evolution in changing environments. In: Monographs in Population Biology, vol. 2. Princeton University Press, Princeton, N.J.

CHAPTER THIRTY ONE

1969 The polygyny threshold model

The concept

Many birds are monogamous but some species exhibit polygynous mating systems. Gordon Orians (1969) developed early ideas about bird mating systems into a formal algorithm to predict when a female might accept a polygynous pairing rather than a monogamous pairing. This algorithm incorporated aspects of the quality of male territory and is called the "polygyny threshold model".

The explanation

Orians (1969) suggests that mate choice may evolve if (i) acceptance of one mate precludes mating with another and (ii) rejecting a mate will likely be followed quickly with another choice of mate. Males that control access to nesting sites may vary in the quality of these sites. Thus, one aspect of female choice among birds would be the relative quality of the sites controlled by different males. However, if a male has already paired with a female then a second possible mate must take into account (i) the reduced time the male may have for care of the young and (ii) the increased demand for resources on a site with two nests.

Orians argues that if two males have identical quality sites but one male is paired and one is not, then a female should always prefer the unpaired male. However, it is possible that if the nest site quality is sufficiently different, then a paired male on a high quality site may be preferred to an unpaired male on an inferior site. The difference in environmental quality required to make such a decision possible is called the polygyny threshold. The term was first used by Verner and Willson (1966). Orians (1969) expands upon the ideas of Verner and Willson and makes seven different predictions from his model, "(1) polyandry should be rare, (2) polygyny should be more common among mammals than among birds, (3) polygyny should be more prevalent among precocial than among altricial birds, (4) conditions for polygyny should be met in marshes more regularly than among terrestrial

Conceptual Breakthroughs in Evolutionary Ecology
ISBN: 978-0-12-816013-8
https://doi.org/10.1016/B978-0-12-816013-8.00031-4

environments, (5) polygyny should be more prevalent among species of early successional habitats, (6) polygyny should be more prevalent among species in which feeding areas are widespread but nesting sites are restricted, and (7) polygyny should evolve more readily among species in which clutch size is strongly influenced by factors other than the ability of the adults to provide food for the young".

Gathering empirical support for this theory was slow but was first provided by Pribil and Searcy (2001). They studied red-winged blackbirds in Ontario, Canada. They argued the polygyny threshold model required that there be evidence that (i) females paid a price for polygyny, (ii) female choice is influenced by male or territory quality, and (iii) if offered a superior male or territory a female would mate polygynously even though unpaired males were available.

Condition (i) was demonstrated by Pribil (2000) by manipulating harem size and showing that females in harems of size 1 had higher reproductive success than those in harems of two. Conditions (ii) and (iii) were tested Pribil and Searcy (2001). In this study female red-winged blackbirds were offered nesting platforms over land with unmated males and nesting sites over water with mated males. The sites over water are consider superior because of the greatly reduced mortality of young in these sites. Pribil and Searcy (2001) observed a strong preference for the superior quality sites despite the presence of a mated male. Pribil (2000) and Pribil and Searcy (2001) provide strong support for the polygyny threshold model, at least in this population of red-winged blackbirds.

Impact: 7

The polygyny threshold model offered a simple evolutionary basis for polygynous mating systems and for the variation seen in mating status of males. Despite the theoretical appeal of this model it is worth noting the 30-year interval between the original proposal and the design and execution of a critical experiment.

References

Orians, G., 1969. On the evolution of mating systems in birds and mammals. Am. Nat. 103, 589–603.

Pribil, S., 2000. Experimental evidence for the cost of polygyny in the red-winged blackbird *Agelaius phoeniceus*. Beyond Behav. 137, 1153–1173.

Pribil, S., Searcy, W.A., 2001. Experimental confirmation of the polygyny threshold model for red-winged blackbirds. Proc. R. Soc. Lond. B 268, 1643–1646.

Verner, J., Willson, M., 1966. The influence of habitats on mating systems of North American passerine birds. Ecology 47, 143–147.

CHAPTER THIRTY TWO

1970 Reproductive effort and the life history schedule

The concept

Gadgil and Bossert (1970) developed a model of life history evolution which explicitly examined age-specific survival and fertility. From this Gadgil and Bossert made predictions about the evolution of iteroparity and semelparity based trade-offs between current and future survival and reproduction.

The explanation

Gadgil and Bossert (1970) sought to study life-history evolution by taking into account many of the ecological variables faced by real populations. This goal eventually led to a model far too complicated to yield analytic results, so instead they relied on computer simulations. Another feature of their model was to follow age-specific reproductive effort. That is the fraction of an organism's energy budget at any age devoted to reproduction. Reproductive effort was in fact a quantity that could be measured in natural populations and thus this model had the appeal of making predictions which could be tested.

The model started with the standard Lotka-Euler equation, $1 = \sum_{x=1}^{d} e^{mx} l_x b_x$. Typically, the survival function, l_x, and the fertility function, b_x, are assumed to be constants. However, Gadgil and Bossert make these demographic parameters functions of the environment and physiological states. Survival is not only a function of age, but also depends on reproductive effort, and the environment through a parameter called degree of satisfaction which ranges from 0 to 1. In addition, predation was modeled as an age-specific constant reflecting the chance of avoiding predation and hence death. Births were likewise dependent on age, reproductive effort, size, and degree of satisfaction. Adults in this model could grow and larger individuals were assumed to produce more offspring. The growth

Conceptual Breakthroughs in Evolutionary Ecology
ISBN: 978-0-12-816013-8
https://doi.org/10.1016/B978-0-12-816013-8.00032-6

of an individual was also a function of age, reproductive effort and degree of satisfaction.

Fitness would be measured as *m* from the Lotka-Euler equation. But the important component of this theory was building trade-offs between reproduction and survival. Gadgil and Bossert compared two functions, a profit and cost function. The profit function measures the contribution to fitness at age-*i* for a specified reproductive effort. This profit function increases with increasing reproductive effort. If this increase has a convex shape, then the optimal reproductive effort can be between 0 and 1 and the life history will most likely be iteroparous. The cost function measures the survival and size reductions at all ages following a reproductive effort at age-*i*. Again, the cost function increases with increasing effort since a large reproductive effort at one age is expected to be followed by a reduced ability to survive and reproduce at later ages. A concave cost function will also favor iteroparity. If the profit function is not convex and the cost function is not concave, then the life history will be semelparous.

Of course, the shape of these cost and profit functions is tied up in the details of the organismal biology. However, as this theory demonstrates those trade-offs have important consequences for the evolution of life-histories.

Impact: 7

Gadgil and Bossert's work was path breaking at the time it was published. They took the standard demographic theory and added the effects of resource variation, variation in reproductive effort, and trade-offs. The ideas in this paper inspired subsequent research in life history evolution.

Reference

Gadgil, M., Bossert, W.H., 1970. Life historical consequences of natural selection. Am. Nat. 104, 1–24.

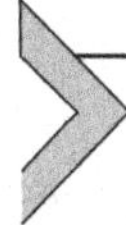

CHAPTER THIRTY THREE

1970 The price equation

The concept

George Price (1970, 1972) showed how the outcome of evolution could be determined by examining the correlation between fitness and phenotypes.

The explanation

The Price equation (Price, 1970, 1972) partitions evolutionary change into two components: (i) a component that reflects changes due to natural selection; and (ii) a second component encompassing all other evolutionary changes. The first component has been the subject of much discussion and some controversy. A particularly insightful review of these issues is given by Frank (2012). The Price equation examines evolutionary change from an ancestral population to a descendant population. The ancestral population is composed of different types each with a frequency, q_i, and fitness value, w_i, for type-i. Associated with each type in the ancestral population is a measured value, z_i, often a phenotype (but see Frank (2012) for a more general discussion of the definition of these quantities). Natural selection will typically result in a change ($\Delta\overline{z}$) in these measured quantities from the ancestral to the descendant population. The changes due to natural selection can be summarized by the first part of the Price equation as $\overline{w}\Delta\overline{z} = Cov(w, z)$, where the bars indicate population averages. The Price equation has been used to study a wide variety of problems including evolution of social behavior (Hamilton, 1970), loss of biodiversity in an ecosystem (Fox, 2006), multilevel selection (Wade, 1985), kin selection (Uyenoyama et al., 1981) and other problems.

Impact: 10

Despite 50 years of earlier population genetic research Price managed to derive, for the first time, a simple and elegant mathematical description of

Conceptual Breakthroughs in Evolutionary Ecology
ISBN: 978-0-12-816013-8
https://doi.org/10.1016/B978-0-12-816013-8.00033-8

the evolutionary process. His description opened up study in areas previously deemed too difficult and as a result has had lasting impact.

References

Fox, J.W., 2006. Using the Price equation to partition the effects of biodiversity loss on ecosystem function. Ecology 87, 2687–2696.

Frank, S.A., 2012. Natural selection. IV. The Price equation. J. Evol. Biol. 25, 1002–1019.

Hamilton, W.D., 1970. Selfish and spiteful behavior in an evolutionary model. Nature 228, 1218–1220.

Price, G.R., 1970. Selection and covariance. Nature 227, 520–521.

Price, G.R., 1972. Extension of covariance selection mathematics. Ann. Hum. Genet. 35, 485–490.

Uyenoyama, M.K., Feldman, M.W., Mueller, L.D., 1981. Population genetic theory of kin selection: I. Multiple alleles at one locus. Proc. Natl. Acad. Sci. U.S.A. 78, 5036–5040.

Wade, M.J., 1985. Hard selection, kin selection, and group selection. Am. Nat. 125, 61–73.

CHAPTER THIRTY FOUR

1971 Population genetic theory of density-dependent natural selection

The concept

MacArthur's ideas about *r*- and *K*-selection, as he called it were only semi-quantitative (MacArthur, 1962; MacArthur and Wilson, 1967). While the logistic equation was referenced in MacArthur's 1962 paper, his 1967 book with E.O. Wilson developed verbal theories of *r*- and *K*-selection. This verbal theory discusses repeated reproduction as a *K*-selected strategy even though the logistic model has no age-structure (Pianka, 1970, 1972). Roughgarden (1971) developed explicit single locus population genetic models and then made predictions for the outcome of evolution at high and low densities which could then be subject to experimental tests.

The explanation

Roughgarden (1971) developed an explicit genetic model in which he assumed fitness was equivalent to per-capita rates of population growth. Accordingly, when growth rates are assumed to be density dependent and they decline linearly with population size, the fitness of genotype A_iA_j is $1+r_{ij}-(r_{ij}N/K_{ij})$, where r_{ij} and K_{ij} are the genotype specific intrinsic rate of increase and carrying capacity of the logistic equation. Roughgarden goes on to analyze the outcome of evolution in stable environments that allow the population to approach carrying capacity. In that case she shows the genotype with the highest K is favored. Roughgarden also analyzes seasonal environments that keep the population size below carrying capacity. In that case, depending on the severity of the environment, the genotype with the highest r can be favored. A detailed stability analysis of these models was also carried out by Charlesworth (1971).

Roughgarden suggests that a possible test of this theory would be to compare the r and K values of *Drosophila melanogaster* taken from low and high latitude environments, the reasoning being that in higher latitudes

Conceptual Breakthroughs in Evolutionary Ecology
ISBN: 978-0-12-816013-8
https://doi.org/10.1016/B978-0-12-816013-8.00034-X

there is more seasonality (and hence higher *r*-values should be at a premium). To my knowledge those measurements have never been made. However, environments with and without extreme "seasonality" were created in the lab specifically to test these ideas (Mueller and Ayala, 1981, Chapter 46).

Impact: 8

Roughgarden's work was the first to link the standard models of population genetics with ecological theory. This work then marks the true union of evolutionary and ecological theory.

References

Charlesworth, B., 1971. Selection in density-regulated environments. Ecology 52, 469–474.
MacArthur, R.H., 1962. Some generalized theorems of natural selection. Proc. Natl. Acad. Sci. U.S.A. 48, 1893–1897.
MacArthur, R.H., Wilson, E.O., 1967. Island Biogeography. Princeton Univ. Press, Princeton.
Mueller, L.D., Ayala, F.J., 1981. Trade-off between *r*-selection and *K*-selection in *Drosophila* populations. Proc. Natl. Acad. Sci. U.S.A. 78, 1303–1305.
Pianka, E.R., 1970. On *r* and *K* selection. Am. Nat. 104, 592–597.
Pianka, E.R., 1972. *r* and *K* selection or *b* and *d* selection? Am. Nat. 106, 581–588.
Roughgarden, J., 1971. Density-dependent natural selection. Ecology 52, 453–468.

1971 The evolution of heavy metal tolerance

The concept

Human mining activities result in the deposition of large quantities of heavy metals at levels lethal to most plants. Nevertheless, over time resistant plants are found in these areas within very short distances from susceptible individuals. These studies illustrate the ability of plant populations to adapt rapidly to changing environmental conditions. These studies also revealed conditions that might lead to reproductive isolation and an increase in self-fertility.

The explanation

In this book I cover two breakthroughs representing studies of adaptation to man-made environmental changes. In Chapter 37 I discuss the impact of pollution from coal burning industry in England starting in the 19th century. In this Chapter I review studies of plant adaptation to heavy metals deposited from mine tailings. Human alteration of the environment is nothing new. However, we are now faced with the well documented increase in global temperatures which are certain to have widespread impact on plant and animal species. Understanding the ways in which organisms adapt to these environmental insults is clearly an important topic.

Although mine tailings are a substantial and obvious mechanism for the rapid deposition of large quantities of heavy metals, other such mechanisms exist. Until recently the gasoline used in automobiles contained lead. Some of this lead is released in the exhaust and often deposited and eventually accumulated along the roadside (Antonovics et al., 1971). Heavy metals have also been used in fungicides, pesticides, and disinfectants. The use of these agents can also result in the environmental accumulation of heavy metals, but then target organisms of the pesticide or fungicide may evolve resistance.

There are several well documented cases of adaptation to heavy metals (Antonovics et al., 1971). Here the focus is on adaptation of two common grass species: *Agostis tenuis* and *Anthoxanthum odoratum* (McNeilly, 1968;

Conceptual Breakthroughs in Evolutionary Ecology
ISBN: 978-0-12-816013-8
https://doi.org/10.1016/B978-0-12-816013-8.00035-1

McNeilly and Antonovics, 1968). The *A. tenuis* population is found near a copper mine while the *A. odoratum* population is near a zinc-lead mine. Plants in the contaminated areas showed high levels of resistance relative to individuals in a nearby pastures less than 100 m away (Jain and Bradshaw, 1966). One of the perplexing issues is how there can be a sharp boundary between resistant and non-resistant individuals over such a short distance.

For instance, why don't the resistant genotypes spread into the nearby pastures? The resistant genotypes don't spread into the uncontaminated pastures a short distance away because they are competitively inferior to the non-resistant genotypes. However, there is evidence of gene flow between the populations. Seed from plants on contaminated sites, when raised and tested under controlled conditions, show lower tolerance than the adults producing those seeds. This suggests that adults in the contaminated areas are receiving pollen from individuals in the non-contaminated pastures. Thus, seeds carry the non-resistant genes and these individuals mostly die and do not survive into the adult population. Conversely, seeds in the pasture areas have greater tolerance than their parents suggesting pollen flow from the contaminated sites (Antonovics et al., 1971). However, selection is so strong that a cline in tolerance appears to be stably maintained.

Given that gene flow results in maladaptive individuals, it would seem that selection should favor a reduction in gene flow. Antonovics (1968) showed no barrier exists to cross-fertilization between individuals in the pasture and contaminated sites. However, *A. odoratum* (and to a lesser extent *A. tenuis*) showed higher rates of self-fertilization in the contaminated areas. Increasing self-fertilization would certainly reduce gene flow as well as cause inbreeding depression. In addition to these fitness related traits, Antonovics et al. (1971) note that there are differences in morphology and physiology of the tolerant and no-tolerant individuals.

Impact: 8

The work on adaptation to heavy metals by Antonovics and others has shown how a cascade of events can result from environmental disturbance.

References

Antonovics, J., 1968. Evolution in closely adjacent plant populations. Heredity 23, 219–238.

Antonovics, J., Bradshaw, A.D., Turner, R.G., 1971. Heavy metal tolerance in plants. Adv. Ecol. Res. 7, 1–85.

Jain, S.K., Bradshaw, A.D., 1966. Evolutionary divergence among adjacent plant populations. I. The evidence and its theoretical analysis. Heredity 21, 407–441.

McNeilly, T., 1968. Evolution in closely adjacent plant populations. III. Agrostis tenuis on a small copper mine. Heredity 23, 99–108.
McNeilly, T., Antonovics, J., 1968. Evolution in closely adjacent plant populations. Heredity 23, 205–218.

CHAPTER THIRTY SIX

1973 The red queen hypothesis

The concept

Motivated by observations of extinction rates in the fossil record, Leigh Van Valen (1973) came up with a high-level theory of evolution he called the Red Queen hypothesis. This theory was designed to explain evolution of interacting species in a common environment. In this theory Van Valen argues that the beneficial evolutionary changes in one species will have negative consequences for other species. Accordingly, those negatively impacted species will respond with their own evolutionary changes which, to some extent, will negate the previous beneficial changes. This will lead to a stochastically constant flux in the evolutionary process that at some levels, like the extinction rates of taxa, will suggest constancy over sufficiently long periods of time.

The explanation

Van Valen starts his theoretical development with a deep survey of extinction rates inferred from the fossil record. If similar taxonomic groups are analyzed together, then one can use the duration of each member of the group in the fossil record to create a sort of survival plot — e.g. how many members of the group survived 10 million years, 20 million years, and so on. On a log scale many of the plots show a linear decline in survival. This suggests a constant rate of extinction. Nothing in real life is simple, so in fact there are a number of groups where a linear decline is not seen. Van Valen devotes a section of his paper subtitled "Apparent Exceptions" to explain away these inconvenient observations. However, in the end Van Valen makes a more sober assessment of the data by saying that nevertheless "There is a strong first-order effect of linearity, and it is this, rather than its perturbations by special and diverse circumstances, that deserves primary attention" (Van Valen, 1973). While this seems like a reasonable way to proceed, one wonders if even a successful explanation of first-order effects deserves the attribution of a law.

Conceptual Breakthroughs in Evolutionary Ecology
ISBN: 978-0-12-816013-8
https://doi.org/10.1016/B978-0-12-816013-8.00036-3

The theory assumes that species like predators or parasites that evolve an increased ability to secure their food will be met with an equal response by their prey to resist those advances. In the end the net change in fitness over time is expected to be 0. Van Valen notes that the law as he developed it cannot be derived from lower level principles, e.g. population genetic models. However, he leaves open the possibility that perhaps he has not been cleaver enough to ferret out these explanations.

Given the prediction that fitness will show no net change over time has led Van Valen to a catchy description of his theory which borrows from the Red Queen's quote in Lewis Carroll's book "Through the looking glass", "Now here you see, it takes all the running you can do, to keep in the same place". The process of evolutionary response by interacting species given by Van Valen is not dissimilar to Ehrlich and Raven's descriptions in their work (Ehrlich and Raven, 1964) on coevolution which is not cited by Van Valen. While the detailed equations of the Red Queen hypothesis have not been widely applied, the broad concept of evolution as a zero-sum game have appeared often in evolutionary discussions.

Impact: 5

The inability of Van Valen to connect the Red Queen hypothesis to basic population genetic theory has probably limited the impact of this theory. However, these ideas set up the notion that coevolving species like parasites and their hosts may have a perpetual arms race. These ideas have been important therefore in setting up a simple expectation for these types of coevolutionary settings.

References

Ehrlich, P.R., Raven, P.H., 1964. Butterflies and plants: a study in coevolution. Evolution 18, 586–608.

Van Valen, L., 1973. A new evolutionary law. Evo. Evol. Theory 1, 1–30.

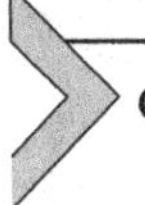

CHAPTER THIRTY SEVEN

1973 The evolution of melanism

The concept

The study of industrial melanism took place over decades, but I mark the event at 1973 (which coincides with the release of the book by Kettlewell reviewing the phenomenon). Industrial melanism is one of the earliest and most studied examples of natural selection in a human-altered environment, and it appears in almost all evolution textbooks as an example of adaptation to environmental change.

The explanation

The peppered moth, *Biston betularia* has a single genetic locus which control the distribution of melanin in the moths' wings. Moths homozygous for the *typica* allele (*c*) are well camouflaged on lichen found on the trunks of trees. Homozygotes for the *carbonaria* allele (*C*) are black. During the 19th century, England underwent dramatic industrialization that was largely fueled by coal-burning plants, the pollution from which would eventually settle on tree trunks. As the effects of pollution spread, the *carbonaria* allele increased in frequency from the time of its first appearance in 1848.

By 1896, the coloration of the peppered moth was viewed as a means of protecting the moth from predators. The *typica* homozygotes would no longer be camouflaged on soot-covered trees. Experiments with predators generally confirmed the advantages of crypsis for the *typica* and *carbonaria* morphs on the different backgrounds (Kettlewell, 1973). Brakefield (1987) discusses some of the more detailed experiments on visual selection of the predators and habitat selection of male and female peppered moths. The outcome of these interactions with predators was that close to the source of pollution the *carbonaria* allele has an advantage (Fig. 37.1). Trees more distant from the city centers have reduced levels of pollution and the *typica* allele is favored (Fig. 37.1).

A common way to study the genetic mechanisms underlying evolutionary change is to use reverse selection (Teótonio and Rose, 2001), usually limited to laboratory populations. Human activity led to the widespread use

Conceptual Breakthroughs in Evolutionary Ecology
ISBN: 978-0-12-816013-8
https://doi.org/10.1016/B978-0-12-816013-8.00037-5

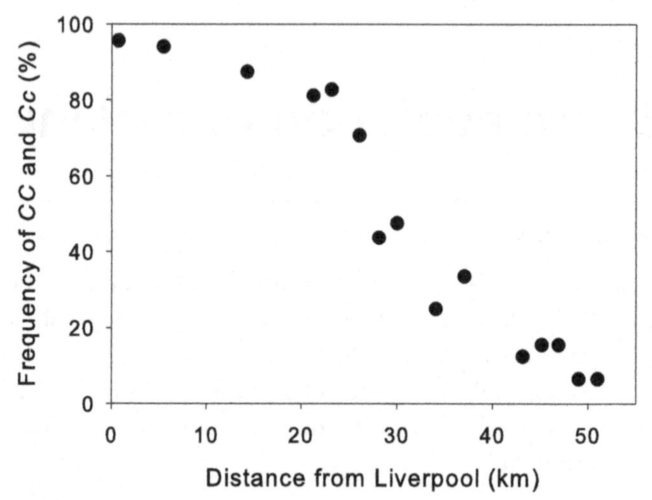

Fig. 37.1 The frequency of *carbonaria* morphs, *CC* and *Cc*, as a function of distance from Liverpool. *(from Mani, 1980).*

of coal as a fuel but eventually the use of coal declined as its harmful health effects were recognized. Between 1960 and the early 1980's, as a result of Clean Air Acts in the United Kingdom, smoke and SO_2 levels dropped substantially. Clarke et al. (1985) showed that a decline in the frequency of the *carbonaria* allele accompanied this decline in pollution.

Impact: 9

The natural experiment created by the rise and decline in coal usage provided a detailed example of adaptation to a changing environment.

References

Brakefield, P.M., 1987. Industrial melanism: do we have the answers? Tree 2, 117–122.

Clarke, C.A., Mani, G.S., Wynne, G., 1985. Evolution in reverse: clean air and the peppered moth. Biol. J. Linn. Soc. 26, 189–199.

Kettlewell, H.B.D., 1973. The Evolution of Melanism. Clarendon Press, Oxford.

Mani, G.S., 1980. A theoretical study of morph ratio clines with special reference to melanism in moths. Proc. R. Soc. London, Ser. A or B 210, 299–316.

Teótonio, H., Rose, M.R., 2001. Perspective: reverse evolution. Evolution 55, 653–660.

CHAPTER THIRTY EIGHT

1975 Sex and evolution

The concept

George Williams (1975) confronted the long-standing problem of why organisms should reproduce sexually given the 50% fitness cost of sexuality. Williams does not believe there is one single answer but many which depend on organismal specifics like the mating system, relative levels of competition and other special life history features. Williams (1975) lays out four different general models applicable to a wide variety of organisms.

The explanation

Compare a diploid organism that reproduces by parthenogenesis to a diploid sexually reproducing organism. The progeny of the asexual individual receives their entire genome from their mother whereas the progeny of the sexually reproducing individual receive only 50% of the genome from their mother. This 50% disadvantage of sex is what Williams aimed to address in his book, Sex and Evolution (Williams, 1975). Williams rejected the general argument that sexual reproduction allows the species to evolve more rapidly. Williams sought explanations that provided an immediate advantage to sex thereby making it more likely that sex would spread in a population as a mode of reproduction. He considered four categories of reproduction which he thought would be particularly conducive to sexual reproduction. I review three of these next.

Aphid-Rotifer model: in this situation Williams posits that an organism occupies small, discontinuous habitats like small pools of water. Once a habitat is occupied the organism may go through many generations of asexual reproduction. Ultimately it will need to disperse propagules to new habitat. At this point Williams compares asexual genotypes which disperse genetically identical clones of the parent to sexual genotypes which produce numerous genetically different propagules. In the new habitats if there are enough generations of asexual reproduction, the sexual genotype is more likely to produce the most superior propagule which will become the

Conceptual Breakthroughs in Evolutionary Ecology
ISBN: 978-0-12-816013-8
https://doi.org/10.1016/B978-0-12-816013-8.00038-7

dominant type in the population and may easily overcome the two-fold disadvantage of sex.

Strawberry-coral model: Williams next considered organisms that are sessile but that grow vegetatively so that they occupy a large space. This clonal growth is eventually halted due either to unsuitability of the environment on the boundary of the clone or the organism meets up with a competing clone. At this point a new clone may drop down in the resident's space and if it is a superior competitor it may displace that resident clone. Over time we expect the resident clones to be well adapted to their local environment. However, the only reasonable way they might expect to expand to other habitats which may also have highly adapted clones is by producing a variety of genotypes through sexual reproduction. If an individual clone has become large through vegetative reproduction it will be able to produce many sexual propagules and hence increase its chance of finding a suitable new territory.

Elm-Oyster model: This model deals with organisms that are sessile as adults but are capable of producing enormous numbers of seeds or larvae. Vast numbers of these seeds will settle in an area capable of supporting perhaps one adult. Thus, there will be intense competition and the use of many asexual clones in this process is doomed to failure. So, unlike the aphid-rotifer model these organisms do not ever reproduce asexually.

Williams (1975) also reviews how sex would benefit mobile animals. In addition to these basic ideas Williams reviews the potential strength of natural selection in organisms with high fecundity.

Impact: 8

Williams' (1975) highly cited work is a masterful combination of evolutionary theory and ecological details in the tradition of Darwin.

Reference

Williams, G.C., 1975. Sex and Evolution. Monographs in Population Biology, vol. 8. Princeton.

CHAPTER THIRTY NINE

1975 Genetics of mimicry

The concept

Clarke and Sheppard (1971) suggested that mimicry in butterflies may be controlled by supergenes that undergo little recombination. Charlesworth and Charlesworth (1975, 1976a,b) developed theoretical population genetic models to show the evolution of such supergenes was possible. Joron et al. (2011) describe such a supergene family in the butterfly *Heliconius numata* which is held together inside a chromosome inversion.

The explanation

Clarke and Sheppard (1971) studied a variety of mimics from Southeast Asian populations of *Papilio memnon*. This species has several different forms that vary in wing patterns and coloration. They also mimic different models. Clarke and Sheppard carried out classical crossing studies and proposed that there were five loci that control presence of tails, hindwing patterns, forewing patterns, color of a forewing area, and abdomen color. They suggested that these genes might be tightly linked in a "supergene" family. This last suggestion was based on the relative rarity of recombinants between these five genes.

Debra and Brian Charlesworth would explore this suggestion through the creation of theoretical population genetic models (Charlesworth and Charlesworth (1975, 1976a,b). Charlesworth and Charlesworth (1975) first consider a single locus model with age-structure to match the life history of the typical mimetic butterfly, although the variables of age-specificity are not explored. This single locus changes the appearance of the mimic to make it resemble a distasteful model to various degrees. Charlesworth and Charlesworth show that unless the original form of the butterfly is very conspicuous, a new allele causing a change in the appearance of the mimic must be a large change resulting in good resemblance of the model. Thus, as others had speculated, this model suggests mimicry will not evolve by small steps but will require rather larger changes. If a new mimetic form is only slightly more conspicuous than the original non-mimetic form, the mimetic allele may sweep to fixation even if the population size of the mimetic species is greater than that of the

Conceptual Breakthroughs in Evolutionary Ecology
ISBN: 978-0-12-816013-8
https://doi.org/10.1016/B978-0-12-816013-8.00039-9

model. Thus, Fisher's original conjecture that the mimetic species population must always be smaller than the model species is not always true.

To explore Clarke and Sheppard's suggestion about supergenes, Charlesworth and Charlesworth (1976a) explored a two-locus model. One locus controlled the basic pattern of mimicry to a model species. The second locus modified the pattern of mimicry by either improving the degree of resemblance or allowing modifications that matched a second model. This second locus could also have effects on non-mimetic individuals making them more conspicuous. Lastly, Charlesworth and Charlesworth looked at the fate of a modifier allele that would reduce the level of recombination between the two loci. This part of the model would explore the idea that evolution might favor a supergene.

Their model showed that once an allele with a major effect has been established at the first locus modifiers with small effects that improve the resemblance may accumulate over time at the second locus. They also show that if the two loci are already linked that there will be selection, albeit weak, to decrease the level of recombination between the loci.

We now go forward 35 years to the work of Joron et al. (2011) on the genetic control of mimicry in the butterfly *Heliconius numata.* This species has a large number of different morphs that are mimics of seven different model species. Joron et al. (2011) found that these mimetic patterns were controlled by a supergene called locus *P*. Using modern molecular techniques Joron et al. found that the *P* locus is associated with a 400-kilobase region inside a chromosome inversion containing 18 different genes. This inversion effectively prevents recombinant genotypes from being created and effectively makes this region a supergene following the predictions of Charlesworth and Charlesworth.

Impact: 9

This work lays out a detailed evolutionary and population genetic story for the evolution of species interactions in the form of Batesian mimicry. It shows the power of evolutionary theory and modern molecular techniques for advancing our understanding of the evolutionary process.

References

Charlesworth, D., Charlesworth, B., 1975. Theoretical genetics of Batesian mimicry I. Single-locus models. J. Theor. Biol. 55, 283–303.

Charlesworth, D., Charlesworth, B., 1976a. Theoretical genetics of Batesian mimicry II. Evolution of supergenes. J. Theor. Biol. 55, 305–324.

Charlesworth, D., Charlesworth, B., 1976b. Theoretical genetics of Batesian mimicry III. Evolution of dominance. J. Theor. Biol. 55, 325–337.

Clarke, C.A., Sheppard, P.M., 1971. Further studies on the genetics of the mimetic butterfly *Papilio memnon* L. Philos. Trans. R. Soc. Lond., B 263, 35–70.

Joron, M., Frezal, L., Jones, R.T., Chamberlain, N.L., Lee, S.F., Haag, C.R., Whibley, A., Becuwe, M., Baxter, S.W., Ferguson, L., Wilkinson, P.A., Salazar, C., Davidson, C., Clark, R., Quail, M.A., Beasley, H., Glithero, R., Lloyd, C., Sims, S., Jones, M.C., Rogers, J., Jiggins, C.D., ffrench-Constant, R.H., 2011. Chromosomal rearrangements maintain a polymorphic supergene controlling butterfly mimicry. Nature 477, 203–206.

CHAPTER FORTY

1976 Evolution of resource partitioning

The concept

The theory of density-dependent natural selection had been applied to the evolution of single populations (Chapter 27). In 1976 Roughgarden expanded this theory to the coevolution of competing species to produce the first comprehensive theory of community evolution.

The explanation

The evolution of parameters of the logistic equation via density-dependent natural selection was reviewed in Chapter 27. Roughgarden (1976) sought to expand this evolutionary model by looking at a community of interacting species. These interacting species might be a predator and prey or it might be a collection of species that compete for a common resource. In either case, the fact that each species may affect the growth of another complicates the evolutionary dynamics. Roughgarden makes some general conclusions about these coevolutionary models before looking at some specific cases.

The example that Roughgarden spends the most time developing is the case of competing species. In the case of two species the per-capita growth rate of species 1 is given by the Lotka-Volterra competition equations as, $1 + r - \frac{rN_1}{K} - \frac{\alpha_{12} r N_2}{K}$, where α_{12} is the competition coefficient which measures the impact of species-2 on the growth of species-1. Competition is for a single resource which can be measured on a single axis, like the length of a seed. The carrying capacity for each species can then be a function (say Gaussian) on the resource axis. Each species has a different location on the resource axis where its carrying capacities is at a maximum. Likewise, competition is assumed to be proportional to the overlap of each species utilization of the resource. With some assumptions, the level of competition between two species may then be a simple function of the distance between the two on the resource axis.

Conceptual Breakthroughs in Evolutionary Ecology
ISBN: 978-0-12-816013-8
https://doi.org/10.1016/B978-0-12-816013-8.00040-5

Roughgarden's model then shows that the evolution of resource use will be a balance between the evolutionary force to reduce competition and the disadvantage to shifting to a new resource type. Roughgarden reaches some general conclusions. If species diversity is not correlated with resource diversity then increasing the number of species in a community will result in neighboring competitors having greater niches overlap. This relationship is not expected if there is a correlation between species and resource diversity.

Impact: 9

Roughgarden's paper was a significant advance in the theory of community evolution. It showed how species interactions, like competition, could affect the evolution of important community properties such as the level of niche overlap.

Reference

Roughgarden, J., 1976. Resource partitioning among competing species-a coevolutionary approach. Theor. Popul. Biol. 9, 388–424.

CHAPTER FORTY ONE

1977 Life history theory challanged

The concept

Despite the development of some substantial theory in evolutionary ecology, the major ideas dominating in the early 1970's were largely verbal or semi-quantitative. Stearns (1977) wrote an insightful review of this field which was instrumental in moving research in life-history evolution to a more quantitative and precise matching of theory with observations.

The explanation

Stearns (1977) identifies what he calls seven ambiguities with theories of life-history evolution. He notes that data or experiments that appear to falsify those theories with ambiguities create a dilemma. The falsification may be due to a central, explicit part of the theory or may be due to one of these ambiguities rendering scientific progress difficult. I summarize these ambiguities next.

(1) *Diploid genetics and ontogeny*. Stearns notes that most genetic models assume single loci while most life history traits are certainly not controlled by a single locus. Stearns also points out that the phenotype may be affected in important ways by the environment rendering a simple one-to-one relationship between phenotype and genotype inaccurate. Finally, Stearns also suggests that regulatory genes may play an important role in life-history evolution.

(2) *Design constraints*. Here Stearns is most concerned with models that deal with determining optimal life-histories. He correctly notes that organisms may have design constraints that prevent them from reaching this optimum. Although Stearns does not discuss this the genetic system may also prevent a population from evolving to the optimum phenotype.

(3) *Multiple causation*. By this Stearns means that there may be multiple theories making the same predictions.

Conceptual Breakthroughs in Evolutionary Ecology
ISBN: 978-0-12-816013-8
https://doi.org/10.1016/B978-0-12-816013-8.00041-7

(4) *Stable age-distribution.* Fitness in age-structured populations is often represented as the rate of exponential growth from the Lotka equation. This calculation assumes that the population is in stable age-distribution (an assumption that is hard to verify in biological populations and is probably never exactly true).

(5) *The carrying capacity as a function of life-history traits.* Simple life history models that keep track of age-specific mortality and fertility can be used to estimate maximum growth rates but they can't provide estimates of the carrying capacity. Stearns argues that simple theories, like the verbal theory of *r*- and *K*-selection that suggest a trade-off between these two logistic parameters can't be connected through a common suite of demographic parameters (although see theories developed since 1977, Mueller, 1988).

(6) *Alternative equilibria.* Stearns notes that some models may have multiple equilibria or models with stochastic effects will need to be summarized by a distribution of results.

(7) *Choice of time scale.* If the life history of different taxa are compared then what time scale is appropriate, absolute or generations?

Impact: 5

While Stearns (1977) did not present any new experimental or theoretical results, his dissection of the existing life-history theory and data analysis was influential for refocusing research in life-history evolution.

References

Mueller, L.D., 1988. Density-dependent population growth and natural selection in food limited environments: the *Drosophila* model. Am. Nat. 132, 786–809.

Stearns, S.C., 1977. The evolution of life history traits: a critique of the theory and a review of the data. Annu. Rev. Ecol. Systemat. 8, 145–171.

CHAPTER FORTY TWO

1977 Natural selection favors reduced variance in fitness

The concept

Gillespie (1977) showed that when offspring numbers for genotypes vary over time (or between individuals), natural selection favors those genotypes with the smallest variance. Thus, adaptation to variable environments will be accomplished both by increasing the number of offspring and reducing the variance in the number of offspring.

The explanation

Gillespie (1977) attacked the problem of environmental variation very differently than Levins (1968). Gillespie (1973, 1974) focused on two sources of variation in offspring production: (i) individuals of a single genotype may vary in offspring production within the same generation due to environmental variation they encounter during their lifetimes; or (ii) offspring production may vary from one generation to the next due to changes in the environment.

When variation is within a generation, fitness is $\mu - \frac{1}{N}\sigma^2$, where μ is the mean number of offspring, N is the population size, and σ^2 is the variance in offspring number (Gillespie, 1974). When offspring numbers vary over time, the geometric mean is the best measure of fitness and that is equal to $\mu - \frac{1}{2}\sigma^2$ (Gillespie, 1973). Gillespie goes on to show that the genotypes with the smallest variance will be favored by natural selection.

These formulations of fitness could lead to the mean number of offspring declining if it was accompanied by a sufficiently large decrease in variance. When there is variance within a generation, the impact of selection on the variance in offspring number is greater in smaller populations. These ideas are also related to the advantage of iteroparity in a variable environment that may produce bad environments capable of wiping out or severely reducing offspring production (see Chapter 29). Iteroparous organisms will have multiple opportunities to reproduce and hence can reduce their variance in offspring production.

Conceptual Breakthroughs in Evolutionary Ecology
ISBN: 978-0-12-816013-8
https://doi.org/10.1016/B978-0-12-816013-8.00042-9

Impact: 10

Gillespie's (1977) simple development of fitness in variable environments led to great evolutionary insights, and led to another of Gillespie's other major accomplishments, which was a model of selection on protein polymorphism (Chapter 43, Gillespie, 1978).

References

Gillespie, J.H., 1973. Natural selection with varying selection coefficients – a haploid model. Genet. Res. 21, 115–120.

Gillespie, J.H., 1974. Natural selection for within-generation variance in offspring number. Genetics 76, 601–606.

Gillespie, J.H., 1977. Natural selection for variances in offspring numbers: a new evolutionary principle. Am. Nat. 111, 1010–1014.

Gillespie, J.H., 1978. A general model to account for enzyme variation in natural populations. V. The SAS-CFF model. Theor. Popul. Biol. 14, 1–45.

Levins, R., 1968. Evolution in changing environments. In: Monographs in Population Biology, vol. 2. Princeton University Press, Princeton, N.J.

CHAPTER FORTY THREE

1978 Maintenance of protein polymorphisms in a variable environment

The concept

In the 1960's and early 1970's, evolutionary biologists actively sought explanations for the vast levels of genetic variation observed in natural populations at protein-coding genes. One simple explanation was that most of these genetic variants were equivalent with respect to their fitness effects such that random genetic drift and mutation were responsible for the observed levels of variation. Gillespie (1978) suggested another feasible explanation: perhaps natural selection and fitness variation due to random environmental changes were responsible.

The explanation

The application of gel electrophoresis (Lewontin and Hubby, 1966; appendix) to natural populations uncovered unexpected levels of genetic variation. These observations demanded explanations. One explanation was that the protein variants revealed by electrophoresis were functionally equivalent and hence neutral with respect to fitness. Thus, the primary forces controlling genetic variation were mutation and random genetic drift. This "neutral" theory of evolution had many appealing features (a review of the theory is given by one of the main proponents - Kimura, 1983 – although see Gillespie, 1983, for an alternative view).

Gillespie developed a model based on the biological properties of enzymes which also incorporated the random variation in fitness over time and space. Gillespie and Langley (1974) argued that it is reasonable to assume that the enzyme activity of a heterozygote is likely to be the average of the activity of homozygotes for each of the two alleles (Fig. 43.1). They also assumed that since the function of protein enzymes was to catalyze biochemical reactions, fitness should be related to enzyme activity. However, the

Conceptual Breakthroughs in Evolutionary Ecology
ISBN: 978-0-12-816013-8
https://doi.org/10.1016/B978-0-12-816013-8.00043-0

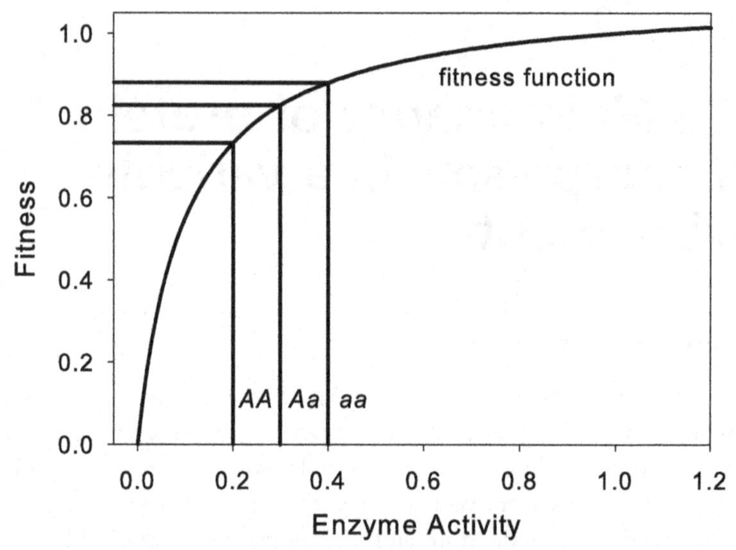

Fig. 43.1 The enzyme activity of three genotypes, *AA*, *Aa*, *aa*, are shown along with their fitness. The enzyme activity of the heterozygote is intermediate to the two homozygotes. The fitness function is concave and translates the enzyme activity into fitness. Due to the concavity of the fitness function, the heterozygotes fitness is always closer to the most fit homozygote.

fitness function relating enzyme activity to fitness is likely to be concave since there will be diminishing returns to increasing activity (Fig. 43.1).

The result of these assumptions is that the fitness of heterozygotes is always closer to the more fit homozygote. Now, suppose that the activity of each homozygotes varies across time according to some random process and on average there is no difference between the activity of either homozygote, e.g. in Fig. 43.1 it is equally likely that the *AA* homozygote would have greater activity than the *aa* homozygote. Over time, the net result is a form of overdominance and the maintenance of both alleles in the population.

Gillespie (1978) called his model the "stochastic additive scale, concave fitness function" model (or SAS-CFF). Gillespie expanded the model to multiple alleles at a single locus, two-loci, and models with population structure. Many of these details are reviewed in Gillespie (1991).

A great advance for the neutral theory of evolution was a sampling theory first described by Ewens (1972). This allowed for a statistical assessment of estimates of allele frequencies derived from samples of natural populations. Much of the strength of the neutral theory drives from its fit to predictions from the sampling theory, after estimates of effective population size (*N*) and mutation rates (μ) are made. Gillespie was able to show that the SAS-CFF model has exactly the same sampling theory, except that the distribution no longer depends on *N* and μ but on the parameters from his selection model.

Impact: 10

Gillespie's research is a tour de force for showing the potential role of selection and environmental variation to mold the properties of genetic variation in natural populations.

References

Ewens, W.J., 1972. The sampling theory of selectively neutral alleles. Theor. Popul. Biol. 3, 87–112.

Gillespie, J.H., Langley, C.H., 1974. A general model to account for enzyme variation in natural populations. Genetics 76, 837–884.

Gillespie, J.H., 1978. A general model to account for enzyme variation in natural populations. V. The SAS-CFF model. Theor. Popul. Biol. 14, 1–45.

Gillespie, J.H., 1983. The status of the neutral theory. Science 224, 732–733.

Gillespie, J.H., 1991. The Causes of Molecular Evolution. Oxford University Press, Oxford.

Kimura, M., 1983. The Neutral Theory of Molecular Evolution. Cambridge University Press, New York.

Lewontin, R.C., Hubby, J.L., 1966. A molecular genetic approach to the study of genic heterozygosity in natural populations. II Amounts of variation and degree of heterozygosity in natural populations of *Drosophila pseudoobscura*. Genetics 54, 595–609.

CHAPTER FORTY FOUR

1979 A critique of the adaptationist program

The concept

Using the architectural wonders of St. Mark's Cathedral in Venice as a metaphor Gould and Lewontin (1979) undertake a detailed critique of the indiscriminate use of adaptationist explanations in evolution, anthropology and behavior. This critique calls for more serious consideration of alternative explanations to adaptation by natural selection.

The explanation

By the late 1970's evolutionary ecology was in full bloom and the use of evolutionary thought had spread to other fields including the new field of sociobiology – the use of biology to explain social behavior. The stated goal of the paper was to counter the appearance of research that viewed "...natural selection as so powerful and the constraints upon it so few that direct production of adaptation through its operation becomes the primary cause of nearly all organic form, function, and behavior." Some (Queller, 1995) saw Gould and Lewontin's work as an attempt to discredit sociobiology. Certainly, sociobiology was one target of Gould and Lewontin but evolutionary ecology also suffered from excessive reliance on adaptationist arguments.

Most importantly Gould and Lewontin made several suggestions for alternative explanations that should be considered in addition to adaptation: 1. *No adaptation and no selection at all.* Genetic differences between populations may be neutral and therefore under the influence of drift only. Deleterious alleles may also rise to high frequency by chance especially in small populations. 2. *No adaptation and no selection on the morphological part at issue* (although this applies to other kinds of phenotypes, not just morphological ones). The form of the part is a correlated consequence of selection directed elsewhere. 3. *The decoupling of selection and adaptation.* Thus, there can be selection without adaptation. For instance, selection will favor increased fecundity even in environments with limited resources.

Conceptual Breakthroughs in Evolutionary Ecology
ISBN: 978-0-12-816013-8
https://doi.org/10.1016/B978-0-12-816013-8.00044-2

Alternatively, there can be adaptation without selection as exemplified by phenotypic plasticity (Chapter 11). 4. *Adaptation and selection but no selective basis for differences among adaptations.* Selection at multiple interacting loci can form multiple local adaptive peaks. Different populations may go to different peaks due to historical accidents that start populations near different domains of attractions. 5. *Adaptation and selection, but the adaptation is a secondary utilization of parts present for reasons of architecture, development or history.*

Impact: 8

Gould and Lewontin's (1979) paper has been very influential in recalibrating the research in evolutionary biology. By late 2019 it had been cited over 8000 times. Their suggestions continue to be relevant 40 years later.

References

Gould, S.J., Lewontin, R.C., 1979. The spandrels of San Marco and the Panglossian paradigm: a critique of the adaptationist programme. Proc. R. Soc. Lond. B 205, 581–598.

Queller, D.C., 1995. The spaniels of St. Marx and the Panglossian paradox: a critque of a rhetorical programme. Quart. Rev. Bio 70, 485–489.

CHAPTER FORTY FIVE

1980 Evolution of philopatry

The concept

Paul Greenwood (1980) reviewed a large body of empirical evidence and showed that birds and mammals show a sex-biased dispersal from their natal habitats. In birds, males are more likely to leave and in mammals females are more likely to leave. Greenwood suggested these patterns were due to evolution and the fitness consequences of dispersal.

The explanation

Dispersal after an animal reaches sexual maturity is an important evolutionary event since this movement represents possible gene flow from one population to another. Gene flow will affect levels of genetic differentiation and local adaptation. Natal dispersal is when juveniles undergo permanent dispersal to another location. Individuals that return to their natal habitat are philopatric.

Greenwood (1980) undertook a large-scale literature review to see if there were discernible trends among birds and mammals. What he observed were sex-biased dispersal patterns that differed in birds and mammals. In birds, females were more likely to undergo natal dispersal whereas in mammals, males were more likely to disperse. Greenwood next developed evolutionary explanations for the bias and differences between birds and mammals.

Greenwood first argued that, generally, dispersal will be beneficial as a means of avoiding inbreeding. Indeed, in the few cases were females disperse in mammals, Clutton-Brock (1989) gathered evidence to suggest that this is done precisely to avoid inbreeding and the deleterious fitness effects that follow from it. But then why not both sexes and why the difference between birds and mammals? Greenwood suggests that the mating system of many birds involves the holding of territory. This Greenwood argues is more easily accomplished by males if they return to their familiar and natal habitat. Greenwood also argues, somewhat less convincingly, that males dispersing to new territory may have trouble mating because they are unfamiliar to the population of females.

Conceptual Breakthroughs in Evolutionary Ecology
ISBN: 978-0-12-816013-8
https://doi.org/10.1016/B978-0-12-816013-8.00045-4

Birds are more frequently monogamous while mammals are more frequently polygynous. Greenwood suggests that in order to both avoid inbreeding and take part in multiple matings, mammalian males are better off dispersing to a new habitat not likely to be frequented by related females. Again, the breeding system of mammals is thought to drive the dispersal pattern.

Greenwood's ideas have stimulated research and 30 years after his seminal review the evolutionary forces at work in dispersing species is still actively debated (Clutton-Brock and Lukas, 2012).

Impact: 6

While this was a review paper, it is still one of the most cited papers in the field and Greenwood's basic explanations of sex-biased dispersal are still relevant.

References

Clutton-Brock, T.H., 1989. Female transfer and inbreeding avoidance in social mammals. Nature 337, 70–72.

Clutton-Brock, T.H., Lukas, D., 2012. The evolution of social philopatry and dispersal in female mammals. Mol. Ecol. 21, 472–492.

Greenwood, P.J., 1980. Mating systems, philopatry and dispersal in birds and mammals. Anim. Behav. 28, 1140–1162.

CHAPTER FORTY SIX

1981 Testing density-dependent naural selection

The concept

Once Roughgarden (1971; Chapter 34) had properly identified density-dependent rates of population growth as the appropriate target of natural selection then experimental tests became possible. Such a test would require an organism that could be raised under different density conditions and one which is amenable to measurements of population growth. Additionally, a species with a short generation time would be preferable. Mueller and Ayala (1981) did such a test with *Drosophila melanogaster*. They found that populations adapted to extreme densities reflected this adaptation in their population growth rates. Most interestingly there was a trade-off. The populations adapted to low density had lower growth rates under crowded conditions than the populations adapted to high density, and vice versa.

The explanation

A genetically variable laboratory population of *Drosophila melanogaster* was divided into three replicate low density, or *r*-selected, populations and three replicate high density or *K*-selected populations. The *r*-selected populations experienced low larval and adult densities while the *K*-selected populations experienced high larval and adult densities. The *r*-selected adult populations sizes were 50 while the *K*-selected were about 1000. After eight generations of adaptation Mueller and Ayala (1981) measured population growth rates at one low (10 adults) and two high densities (750 and 1000 adults).

Mueller and Ayala (1981) found that the *r*-selected populations had higher per-capita growth rates and net productivity (a measure of total offspring produced without regard to the timing) than the *K*-selected populations at a density of 10. However, at the two high densities the *K*-selected populations performed better. These results showed that density-dependent rates of population growth may evolve and that there appears to be a trade-off. Whatever traits enhance growth rates at low density to

Conceptual Breakthroughs in Evolutionary Ecology
ISBN: 978-0-12-816013-8
https://doi.org/10.1016/B978-0-12-816013-8.00046-6

not enhance growth rates under crowded conditions. The importance of trade-offs as first developed by Cody (1966; Chapter 24) were experimentally supported by Mueller and Ayala's work (1981).

The results of Mueller and Ayala (1981) were not without qualifications. First, the increased growth rates of the *r*-selected populations at low density were not statistically significant. Second, the breeding population sizes and thus the effective population size of the *r*-selected and *K*-selected populations were very different making the effects of drift in the two selection regimes different.

A new experiment was conducted by Mueller et al. (1991). In this experiment the three *r*-selected populations were crossed among each other after 198 generations of adaptation to the low-density environment. Three replicates of the crossed population were maintained at low density and three were kept at high density. The breeding size of the low-density populations was also elevated to 500. After 25 generations of adaptation, significant differences in growth rates were observed at both low density ($r > K$) and high density ($K >$ r), consistent with the earlier findings. These populations have also been used to study how adaptation to extreme densities has affected competitive ability (Chapter 55), and several larval behavioral traits (reviewed in Chapter 6 of Mueller and Joshi, 2000).

Impact: 9

Mueller and Ayala (1981) subjected one of the early important theories in evolutionary ecology to a rigorous experimental test. This work supported the utility of framing life-history evolution in terms of trade-offs. It also provided an example of how experimental evolution could be used test important principles of evolutionary ecology.

References

Cody, M., 1966. A general theory of clutch size. Evolution **20**, 174–184.

Mueller, L.D., Ayala, F.J., 1981. Trade-off between *r*-selection and *K*-selection in *Drosophila* populations. Proc. Natl. Acad. Sci. U.S.A. 78, 1303–1305.

Mueller, L.D., Guo, P.Z., Ayala, F.J., 1991. Density-dependent natural selection and trade-offs in life history traits. Science 253, 433–435.

Mueller, L.D., Joshi, A., 2000. Stability in Model Populations. Monographs in Population Biology. Princeton University Press, Princeton, N.J.

Roughgarden, J., 1971. Density-dependent natural selection. Ecology 52, 453–468.

CHAPTER FORTY SEVEN

1981 Evolution of population stability: theory

The concept

In 1974 Robert May showed that relatively simple deterministic population growth models can show a wide array of dynamic behavior, from stable points, to cycles all the way to dynamics with seemingly no pattern called chaos. Ecologists initially studied laboratory populations and determined that the large majority seemed to have stable dynamics in the laboratory (Thomas et al., 1980; Mueller and Ayala, 1981a). This lead Heckel and Roughgarden (1980) and Turelli and Petry (1980) independently to explore whether evolution might naturally favor stable population dynamics.

The explanation

Thomas et al. (1980) examined the population dynamics of 27 different species of *Drosophila* in the laboratory. After concluding that they all showed stable dynamics the authors suggested that this may be due to the action of group selection. That is, if a whole population had chaotic dynamics the large fluctuations characteristic of chaos would make these populations more likely to be driven to extinction when combined with, say, some unfavorable weather or other environmental conditions. Heckel and Roughgarden and Turelli and Petry both sought individual selection explanations for the evolution of stability.

Heckel and Roughgarden (1980) used a discrete time logistic equation to study evolution in a random environment. They assumed that populations were near their equilibrium size but that the carrying capacity of alternative genotypes at a single locus were subject to random variation from one generation to the next. Thus, genotype *AA* at a single locus had a carrying capacity that varied over time as, $K_{AA,t} = \overline{K}_{AA} + k_t$, where k_t is a bounded random variable with zero mean. Alternative genotypes were also assumed to vary in the intrinsic rates of increase (r). Heckel and Roughgarden found that this model resulted in the evolution of smaller values of r and hence

Conceptual Breakthroughs in Evolutionary Ecology
ISBN: 978-0-12-816013-8
https://doi.org/10.1016/B978-0-12-816013-8.00047-8

increased stability since for the logistic equation stability is determined by r (when $r < 2$, K is stable and when $r > 2.57$, dynamics are chaotic).

Turelli and Petry (1980), whose paper was published in the same issue of the *Proceedings of the National Academy of Sciences*, suggested that the evolution of stability depended on model details. Turelli and Petry examined growth models of the general form, $N_{t+1} = N_t G[(N_t/K)^{\theta}]$, where the per-capita growth function G was either linear, exponential or hyperbolic. The stability of the equilibrium, K, in these models is determined by both r and θ. In these models Turelli and Petry introduced random fluctuations in either a density-dependent fashion, by multiplying K by $1+z_t$, or a density-independent fashion by multiplying G by $1+ z_t$.

When only r is allowed to evolve, the outcome of evolution is model dependent. Thus, evolution can lead to stable dynamics or to increasing r until dynamics become chaotic. However, if θ is allowed to evolve then the evolution of stability is found across the models examined. This theory leaves open the hard empirical problem of whether there are individual traits whose changes will alter θ in the desired direction. Mueller and Ayala (1981b) showed that populations of *D. melanogaster* homozygous for different whole second chromosomes varied in their estimated values of θ. However, this result doesn't ensure additive genetic variation for this trait in outbred populations.

Impact: 7

The theory described here tackled the difficult question of the evolution of population stability. These results suggested that the choice of models was an important issue. It also pointed to the need for empirical research since it was unclear whether the evolution of an organism's life history could change in the desired direction to modify the stability of whole populations.

References

Heckel, D., Roughgarden, J., 1980. A species near its equilibrium size in a fluctuating environment can evolve a lower intrinsic rate of increase. Proc. Natl. Acad. Sci. U.S.A. 77, 7497–7500.

May, R.M., 1974. Biological populations with nonoverlapping generations: stable points, stable cycles, and chaos. Science 186, 645–647.

Mueller, L.D., Ayala, F.J., 1981a. Dynamics of single species population growth: stability or chaos? Ecology 62, 1148–1154.

Mueller, L.D., Ayala, F.J., 1981b. Dynamics of single species population growth: experimental and statistical analysis. Theor. Popul. Biol. 20, 101–117.

Thomas, W.R., Pomerantz, M.J., Gilpin, M.E., 1980. Chaos, asymmetric growth and group selection for dynamical stability. Ecology 61, 1312–1320.

Turelli, M., Petry, D., 1980. Density-dependent selection in a random environment: an evolutionary process that can maintain stable population dynamics. Proc. Natl. Acad. Sci. U.S.A. 77, 7501–7505.

CHAPTER FORTY EIGHT

1982 Life history evolution in nature

The concept

The population genetic theory in age-structured populations predicts that "...demographic shifts which tilt the age-structure of the population in favor of younger individuals, or which result in a a more rapid decline in fecundity with age, will cause an increased weighting of sensitivity of fitness toward changes at earlier ages" (Charlesworth, 1994, p. 196). Reznick and Endler (1982) studied a natural river system in which demographic shifts in age-structure were induced by predators that fed on different size-classes (ages) of guppies. The different selection pressures caused by these different predators resulted in life history differences among these guppy populations which was consistent with the theory of selection in age-structured populations (Hamilton, 1966).

The explanation

Guppies, *Poecilia reticulata*, in the streams of Trinidad encounter a variety of predators. *Rivulus hartii* is an omnivore that feeds primarily on small immature guppies. Another predator, *Crenicichla alta*, preferentially preys on large, mature guppies. Female guppies have indeterminant growth and thus their size is roughly correlated with their age. The expectation would then be that the age-structure in a stream dominated by *Crenicichla* would be more heavily weighted with young individuals compared to a stream dominated by *Rivulus*. Accordingly, early reproduction should be at a greater premium in the *Crenicichla*-dominated streams compared to the *Rivulus* streams.

Reznick and Endler tested this prediction by measuring the brood interval, time between successive broods, size of mature males, minimum size of a gravid female, and the reproductive allotment which is the relative allocation of energy to reproduction. In a *Crenicichla* stream, all of these measures should be smaller except for reproductive allocation which should be larger than in *Rivulus* streams. This was in fact what was observed and provided

Conceptual Breakthroughs in Evolutionary Ecology
ISBN: 978-0-12-816013-8
https://doi.org/10.1016/B978-0-12-816013-8.00048-X

evidence of life-history evolution in response to predator-induced differences in age-specific selection intensity.

A further test of this system was conducted by Reznick et al. (1990). In this experiment Reznick and his colleagues took guppies that had long lived in a stream with *Crenicichla* and moved them to a stream containing no guppies but did contain a population of *Rivulus*. After several years the female guppies introduced to the *Rivulus* site increased their size at which they initiated reproduction and decreased their reproductive allocation, all changes consistent with an increase in the proportion of older individuals in the population.

Impact: 9

The Trinidad guppy system provided a unique opportunity to experimentally test some basic theories of life-history evolution. Reznick and his colleagues have thereby added to theoretical and laboratory experimental evidence that makes the current Hamilton- Charlesworth theory of selection in age-structured populations one of the strongest in evolutionary ecology.

References

Charlesworth, B., 1994. Evolution in Age-Structured Populations. Cambridge University Press, Cambridge.

Hamilton, W.D., 1966. The moulding of senescence by natural selection. J. Theor. Biol. 12, 12–45.

Reznick, D., Bryga, H., Endler, J.A., 1990. Experimentally induced life-history evolution in a natural population. Nature 346, 357–359.

Reznick, D., Endler, J.A., 1982. The impact of predation on life history evolution in Trinidadian guppies (*Poecilia reticulata*). Evolution 36, 160–177.

CHAPTER FORTY NINE

1982 Evolution of virulence

The concept

Anderson and May (1982) review several theoretical models of host/parasite evolution and population dynamics. They conclude that the key to understanding the level of parasite virulence is understanding how the life-history of the host and parasite connect virulence and parasite transmission. Those details may favor essentially zero virulence, very high virulence or something in-between. This work helps not only to understand the levels of parasite virulence but also conditions under which an antagonistic relationship may evolve into a mutualistic one.

The explanation

The study of the joint evolution of parasites and their hosts is difficult for several reasons. A major difficulty is the different time scales of reproduction for hosts and their parasites, with parasites almost always reproducing more rapidly than their hosts. Anderson and May (1982) note after reviewing genetic and epidemiological models of parasites and host that many of the dynamical details depend not just on the virulence of the parasite but on a combination of virulence and transmissibility. These two traits are unlikely to be independent of each other, but any relationship is crucial to understanding the dynamics.

Anderson and May focus their discussion on the reproductive rate of the parasite since this will happen on a much faster time-scale than host reproduction and thus is likely to be a focus of evolutionary change. For a simple epidemiological model (Anderson and May, 1981) they note that this reproductive value, R_0, is equal to,

$$R_0 = \frac{\beta N}{(\alpha + b + \nu)}$$

where β is the transmission rate of the parasite, α, is the virulence or chance of the host death due to infection, b is the disease-free host death rate, and ν is the host recovery rate. If virulence was independent of the other

Conceptual Breakthroughs in Evolutionary Ecology
ISBN: 978-0-12-816013-8
https://doi.org/10.1016/B978-0-12-816013-8.00049-1

parameters then α should approach zero. However, Anderson and May show this is unlikely to be the case. As an example, they show that different strains of the myxoma virus that infect rabbits show a correlation between α and ν. Those strains with low virulence have high rates of host recovery. If the rabbits recover too quickly the virus titer will be low when the mosquito vector bites the rabbits, thus lowering the reproductive rate of the virus.

Alternatively, if virulence was not related to recovery and negatively related to transmission, then natural selection should strongly favor low virulence. Anderson and May (1982) accordingly conclude that "These formal studies make it clear that the coevolutionary trajectory followed by any particular host-parasite association will ultimately depend on the way the virulence and the production of transmission stages of the parasite are linked together: depending on the specifics of this linkage, the coevolutionary course can be toward essentially zero virulence, or to very high virulence, or to some intermediate grade".

Impact: 7

This paper was important for focusing attention on the importance on the dual action of virulence and transmission. It would be an important spark for experimental studies like Bull et al. (1991) reviewed in Chapter 58.

References

Anderson, R.M., May, R.M., 1981. The population dynamics of microparasites and their invertebrate hosts. Phil. Trans. Roy. Soc. B 291, 451–524.

Anderson, R.M., May, R.M., 1982. Coevolution of hosts and parasites. Parasitology 85, 411–426.

Bull, J.J., Molineux, I.J., Rice, W.R., 1991. Selection of benevolence in a host-parasite system. Evolution 45, 875–882.

CHAPTER FIFTY

1984 Evolution of age-specific patterns of survival and fecundity

The concept

In Chapter 4 Norton's theory of natural selection in age-structured populations was laid out. That theory would later be used by Haldane (1941), Hamilton (1966), Medawar (1952), Williams (1957), and Charlesworth (1994) to develop theoretical arguments for the consistently observed pattern of increasing mortality rates and declining fecundity with age. What remained to be done was a rigorous experimental test of the evolutionary theory. Such tests were provided by Rose and Charlesworth (1981) and Rose (1984). By propagating populations by only using eggs laid by older females, Rose was able to show that existing genetic variation in the starting populations could substantially extend the longevity of fruit flies and their fecundity later in life. These observations demonstrated that by changing the way natural selection typically acts one could postpone the aging process.

The explanation

In an age-structured population we expect the strength of selection to exponentially decline as a function of age. Consequently, selection is expected to most strongly affect increases in survival and fertility early in life even to the detriment of those traits later in life. Rose decided to culture *Drosophila melanogaster* populations by only collecting eggs for the next generation from flies that could survive to advanced ages. A control was kept in which flies reproduced early in life. Each reproductive schedule was replicated in five independent populations in a large, outbred population previously adapted to the laboratory environment. After 15 generations of such selection the survival and reproduction of the late reproducing and early reproducing lines were compared.

Conceptual Breakthroughs in Evolutionary Ecology
ISBN: 978-0-12-816013-8
https://doi.org/10.1016/B978-0-12-816013-8.00050-8

Longevity increased 30% in the late reproducing females and 15% in males relative to controls. Likewise, fecundity of late reproducing females was depressed when they were young and elevated as they aged relative to controls. Taken together these experimental results show that the patterns of age-specific survival and fecundity are molded by natural selection and natural populations harbor genetic variability which may quickly respond to changes in these patterns of age-specific natural selection.

These experiments were sufficiently important that this type of selection protocol has been replicated in other *D. melanogaster* populations (Luckinbill et al., 1984), mice (Nagai et al., 1995), bean Weevils (Tucić et al., 1997), and houseflies (Reed and Bryant, 2000). These experiments showed conclusively that patterns of age-specific survival and fecundity are subject to modification based on the intensity of selection at different ages. Does this type of selection actually happen in nature? Such an example is reviewed in Chapter 48.

Impact: 9

Rose's experiments showed the flexibility of the aging process. Rather than considering aging as foregone conclusion due to the wearing out of parts, his work demonstrated the important role of age-specific natural selection in shaping these patterns. He also revealed that substantial genetic variation exists that can profoundly change these patterns.

References

Charlesworth, B., 1994. Evolution in Age-Structured Populations, second ed. Cambridge University Press, New York.

Haldane, J.B.S., 1941. New Paths in Genetics. George Allen and Unwin, London.

Hamilton, W.D., 1966. The moulding of senescence by natural selection. J. Theor. Biol. 12, 12–45.

Luckinbill, L.S., Arking, R., Clare, M.J., Cirocco, W.C., Buck, S.A., 1984. Selection for delayed senescence in *Drosophila melanogaster*. Evolution 38, 996–1003.

Medawar, P.W., 1952. An Unsolved Problem in Biology. H. K. Lewis, London.

Nagai, J., Lin, C.Y., Sabour, M.P., 1995. Lines of mice selected for reproductive longevity. Growth Dev. Aging 59, 79–91.

Reed, D.A., Bryant, E.H., 2000. The evolution of senescence under curtailed life span in laboratory populations of *Musca domestica* (the housefly). Heredity 85, 115–121.

Rose, M.R., 1984. Laboratory evolution of postponed senescence in *Drosophila melanogaster*. Evolution 38, 1004–1010.

Rose, M.R., Charlesworth, B., 1981. Genetics of life-history in *Drosophila melanogaster*. II. Exploratory selection experiments. Genetics 97, 187–196.

Tucić, N., Stojković, O., Gliksman, I., Milanović, D., Šešlija, D., 1997. Laboratory evolution of life-history traits in the bean weevil (*Acanthoscelides obtectus*): the effects of density-dependent and age-specific selection. Evolution 51, 1896–1909.

Williams, G.C., 1957. Pleiotropy, natural selection, and the evolution of senescence. Evolution 11, 398–411.

CHAPTER FIFTY ONE

1985 Evolution of phenotypic plasticity theory

The concept

The adaptive qualities of phenotypic plasticity have been well appreciated by evolutionary ecologists. But how would natural selection shape patterns of phenotypic plasticity in the face of different selection pressures in nearby environments and gene flow between these different populations? Via and Lande (1985) tackled this problem with a quantitative genetic evolutionary model. They outlined conditions under which evolution would allow populations to achieve their optimal fitness in two different environments and helped place the evolution of phenotypic plasticity in a modern population genetic context.

The explanation

The impact of differential selection across multiple environments or niches had been studied by population geneticists for some time prior to Via and Lande's work (Levene, 1953). But these single-locus models aimed to explore the conditions under which a genetic polymorphism could be maintained. Certainly, in the 1960's as mounting evidence was collected showing high levels of genetic variation in natural populations, these multiple-niche models became part of the selectionist program for attempting to understand the forces maintaining this variation.

But as a tool for studying phenotypic plasticity these models had limited utility. Most traits that exhibit phenotypic plasticity are more likely to be under the influence of many genes. Phenotypic plasticity is most often characterized by the range of phenotypes called the norm of reaction exhibited in different environments. Models treating the evolutionary dynamics of the norm of reaction did not exist. Via and Lande (1985) developed their theoretical models to directly address these issues.

The Via and Lande model have several key components. It deals with the expression of a single trait in two different environments. The expressions in each environment are called character states and they are treated as

Conceptual Breakthroughs in Evolutionary Ecology
ISBN: 978-0-12-816013-8
https://doi.org/10.1016/B978-0-12-816013-8.00051-X

correlated characters in a standard quantitative genetic model. In the first model, individuals disperse to each environment and selection operates to change the mean breeding value based on the phenotype and selection differential. Each environment may either contribute a constant fraction of survivors to the breeding population (soft selection) or they may contribute a number that is proportional to the mean fitness of the population in each environment (hard selection). A second model looks at selection on populations which reside within each of the two environments but exchange some fraction of the population as migrants each generation.

As long as the genetic correlations are not $\pm$ 1, then evolution will take the populations to their optimal phenotypes in each environment although the trajectories may not be direct. Via and Lande note that at the equilibria, stabilizing selection will eventually erode genetic variability. Via and Lande's model does not account for this. However, the authors assert that variability will be maintain by a selection mutation balance as explored by Turelli (1984).

If genetic variation is exhausted during selection then Via and Lande note this will prevent the population from attaining the optimal phenotype. Also, if the subdivided populations do not exchange any migrants then each separate population will adapt to its local environment but not to the other. Although Via and Lande do not explore this alternative formally, they note that if there is a cost to plasticity then evolution may not proceed to the optimal end point.

Impact: 7

The Via and Lande paper came out at a time when there was heightened interest in measuring quantitative genetic parameters in natural populations. This work gave some guidance as to what might be expected and how to interpret these measurements. Via and Lande's theoretical work marked an achievement in the future study of phenotypic plasticity.

References

Levene, H., 1953. Genetic equilibrium when more than one ecological niche is available. Am. Nat. 87, 311–313.

Turelli, M., 1984. Heritable genetic variation via mutation-selection balance: Lerch's zeta meets the abdominal bristle. Theor. Popul. Biol. 25, 138–193.

Via, S., Lande, R., 1985. Genotype-environment interaction and the evolution of phenotypic plasticity. Evolution 39, 505–522.

CHAPTER FIFTY TWO

1985 Coevolution of bacteria and phage

The concept

The simple prediction of the Red Queen hypothesis (Chapter 36) is that species engaged in antagonistic interactions (such as predators and prey or host and parasites) will engage in an endless arms race fueled by evolution. Lenski and Levin (1985) sought to experimentally test this with laboratory communities of bacteria and phage. They found that evolution is constrained and that endless responses by phage and their host bacteria are not what is observed.

The explanation

Phage require bacteria to reproduce. Infected bacteria are ultimately killed by phage and thus the interaction between bacteria and phage is exclusively antagonistic. Bacteria (*Escherichia coli*) and phage (T2, T4, T5, and T7) reproduce quickly and thus their evolution can be experimentally studied. Bacteria can become resistant to phage through single mutations that prevent the adsorption of phage. Phage accomplish adsorption via the recognition of specific lipopolysaccharides or proteins in the bacteria cell wall. Phage may also experience host-range mutations that allow the phage to infect both susceptible and resistant bacteria.

Mutant bacteria resistant to all four phage were observed. Mutant bacteria resistant to T2, T4, and T7 showed a competitive disadvantage relative to the parental, susceptible bacteria. Mutants resistant to T5 did not show a competitive disadvantage. The cultures with bacteria resistant to T2, T4, and T7 did not drive the phage to extinction. In these cultures, both resistant and susceptible bacteria coexisted with a population of unmutated phage. The competitive disadvantage of the resistant bacteria prevented them from displacing the susceptible bacteria and these susceptible bacteria prevented the phage from being driven to extinction. Cultures with T5 showed different results. Once a resistant bacterial mutant appeared the phage were driven to extinction.

Conceptual Breakthroughs in Evolutionary Ecology
ISBN: 978-0-12-816013-8
https://doi.org/10.1016/B978-0-12-816013-8.00052-1

Lenski and Levin didn't observe any host-range mutants. By estimating an upper bound for the mutation rate of host-range mutants, Lenski and Levin conclude that it would take about 7 years for such a mutant to appear whereas the bacterial mutants appeared in about 100 h. Lenski and Levin suggest that genetic changes that produce a resistant bacterium are much simpler than the changes that would be required on the part of phage to extend their host range. Thus, the lack of an arms race is due to architectural constraints on the evolution of phage despite the strong selective advantage that would accrue to such a mutant. Lenski and Levin (1985) conclude that "This general asymmetry in the coevolutionary potential of bacteria and phage occurs because mutations conferring resistance may arise by either the loss or alteration of gene function, while host-range mutations depend on specific alterations of gene function. These constraints preclude observing endless arms races between bacteria and virulent phage."

Impact: 9

These experiments provided a powerful demonstration of the complications in the coevolutionary process. It is not enough to understand the selective forces in play but also evolutionary constraints and fitness trade-offs (see also Chapter 44).

Reference

Lenski, R.E., Levin, B.R., 1985. Constraints on the coevolution of bacteria and virulent phage: a model, some experiments, and predictions for natural communities. Am. Nat. 125, 585–602.

CHAPTER FIFTY THREE

1986 Evolution of Darwin's finches

The concept

Rosemary and Peter Grant (Grant, 1986; Grant and Grant, 1989) devoted their careers to the study of the ecology and evolution of Darwin's finches. During these studies they have carefully documented ecological causes and phenotypic responses to natural selection in detail seldom available.

The explanation

Darwin was not the first person to see or write about the finches on the Galápagos Islands but he was the first to make a careful study of the habits and bring specimens back to England. These specimens included 9 of the 14 species. It would be left to Lack (1947) to construct the first phylogenetic tree of these 14 species. By the early 1970's the Grants were collecting ecological and genetic data on the finches on a yearly basis. Thus, in the middle of 1976 through early 1978 they documented a large decline in the population size of *Geospiza fortis* on the island Daphne Major. There was a substantial drought in 1977 which contributed to this decline. These events set the stage for the Grant's to witness natural selection on this stressed population.

The simple theory of natural selection on a quantitative traits (a trait under the control of many loci) is that there must be additive genetic variance for the trait. The ratio of the additive genetic variance to the total phenotypic variance is called the heritability and it can take values from 0 to 1. High heritability means that selection on that trait will be accompanied by a genetically based change in the mean value of the trait. Grant (1986) documents measurements of heritability for *G. fortis* for six traits; weight, wing length, tarsus length, bill length, bill depth, and bill width. These heteritabilities were all very high and close to 1. Thus, the conditions existed for selection to change these phenotypes if any of them contributed to

Conceptual Breakthroughs in Evolutionary Ecology
ISBN: 978-0-12-816013-8
https://doi.org/10.1016/B978-0-12-816013-8.00053-3

survival in these new, stressful conditions. Grant (1986) documents strong selection on bill depth during the first and second half of 1977.

What caused the selection? The drought reduced the seed output of plants dramatically. As a result, small seeds were quickly consumed leaving predominantly large, hard seeds mostly from *Opuntia* and *Tribulis*. The Grant's had documented that tearing apart the mericarps (a woody fruit part of the plant) of *Tribulis* are best accomplished with a narrow, deep bill. This advantage undoubtedly contributed to the observed evolution of bill depth.

Impact: 10

The work by the Grant's on Darwin's finches is unparalleled among studies of natural populations. Only by the prolonged observation, genetic analysis, and intimate understanding of the physiology, morphology and ecology of these birds would the detailed picture of selection emerge.

References

Grant, P.R., 1986. Ecology and Evolution of Darwin's Finches. Princeton University Press, Princeton, N.J.

Grant, B.R., Grant, P.R., 1989. Evolutionary Dynamics of a Natural Population: The Large Cactus Finch of the Galápagos. University of Chicago Press, Chicago, Ill.

Lack, D., 1947. Darwin's Finches. Cambridge University Press, Cambridge.

CHAPTER FIFTY FOUR

1986 Evolution across three trophic levels

The concept

Ecological interactions among species in different trophic levels have been a focus of study since Lindeman's pioneering work in (Lindeman, 1942). Weis and Abrahamson studied how one trophic level affects the evolution of other trophic levels. This study system provides a rare and detailed look at evolution among interacting species.

The explanation

Gallmakers are insects that stimulate tumorlike growths on plants. These galls serve as protection and food for the insect larvae. Weis and Abrahamson (1986) and Abrahamson and Weis (1997) studied the goldenrod, *Solidago altissima*, which can be induced to make galls by the parasitic insect *Eurosta solidaginis*. The gallmaker, *E. solidaginis*, can also be a host for parasitic wasps, *Eurytoma gigantea* and *Eurytoma obtusiventris*. These wasps insert their ovipositor through the gall and lay eggs on or in the gallmaker. When the wasp egg hatches it consumes the gallmaker.

If the gall is too large, then the wasp will be unable to penetrate to the gallmaker chamber. However, large galls attract insectivorous birds that can peck open the galls and extract the gallmakers. These observations suggest that there might be selection on the ability of gallmakers to influence the size of galls. To study, this Weis and Abrahamson (1986) took samples of galls after parasite attack and predator attack and determined the size of attacked and unattacked galls and the number of gallmakers emerging from these galls. By comparing the mean size of the galls with surviving gallmakers to the entire distribution, the researchers, were able to detect if there was directional and/or stabilizing selection on gall size. Weis and Abrahamson (1986) found significant directional selection (for increased gall size) and stabilizing selection. Of course, this selection applies to the fitness of the gallmakers. Thus, an important question was the extent to which the gallmaker can influence the size of the plant gall.

Conceptual Breakthroughs in Evolutionary Ecology
ISBN: 978-0-12-816013-8
https://doi.org/10.1016/B978-0-12-816013-8.00054-5

In the greenhouse Weis and Abrahamson (1986) raised full-sib families of gall makers on goldenrod which were clones from the same genotype. Thus, there would be no variation in gall size due to genetic variation in the plant; all the variation would be due to genetic differences among the gallmakers or environmental variation. These experiments showed there was significant heritable variation in gall size due to the gallmaker. Thus, even though the plant is growing the tissue for the gall, the genotype of the gallmaker can affect the size of the gall. Selection due to differential survival of gallmakers based on size of the gall can then affect the differential survival of those genotypes of gallmakers which produce the best size galls. This dynamic suggests that the particular distribution of gall size in goldenrod will be a function of the relative strengths of selection on gallmakers due to wasp parasitoids or bird predation. These pressures are likely to vary geographically. Such patterns are consistent with the geographic mosaic theory of coevolution (Thompson, 1994, Chapter 60).

Impact: 8

The research is one of the most detailed examinations of coevolution among an interacting group of three species. Many more details of this work are given in the detailed monograph by Abrahamson and Weis (1997).

References

Abrahamson, W.G., Weis, A.E., 1997. Evolutionary Ecology across Three Trophic Levels. Princeton University Press, Princeton, N. J.

Lindeman, R.L., 1942. The trophic-dynamic aspect of ecology. Ecology 23, 399–417.

Thompson, J.N., 1994. The Coevolutionary Process. University of Chicago Press, Chicago.

Weis, A.E., Abrahamson, W.G., 1986. Evolution of host-plant manipulation by gall makers: ecological and genetic factors in the *Solidago-Eurosta* system. Am. Nat. 127, 681–695.

CHAPTER FIFTY FIVE

1988 Evolution of competitive ability

The concept

The expansive elaboration of *r*- and K-selection by MacArthur and Wilson (1967) predicted that, at high population densities, competitive ability would be at a premium and under strong natural selection for improvement. This prediction was tested experimentally by Mueller (1988a) by keeping replicated populations of *Drosophila melanogaster* at high and low population sizes. After 128 generations of evolution at these extreme densities, the populations adapted to high densities had competition coefficients that were 58% greater than the populations kept at low density. This finding was consistent with basic models of density-dependent natural selection.

The explanation

Mueller (1988a) initiated an experiment to examine the outcome of density-dependent natural selection by using replicated populations of *Drosophila melanogaster*. In Chapter 35 a trade-off in density-dependent population growth rates was described (Mueller and Ayala, 1981). The finding that *Drosophila* had adapted to different levels of crowding motivated further investigation on competitive ability.

In these laboratory-selected populations, crowding has the largest impact on the mortality of the larvae. *Drosophila* larvae compete through a scramble competition for food. Consequently, a good measure of competitive ability is to allow larvae to compete for carefully controlled amounts of food (live yeast). Phenotypically the populations adapted to low and high densities were wild type. Thus, each of these different density-adapted populations was competed against a standard population carrying the X-linked *white* allele, which results in flies with white eyes, at 10 different yeast levels.

The survival of the lab-adapted larvae relative to the white competition stock can be used to infer competitive ability from models of *Drosophila* competition (Mueller, 1988b). The average high-density adapted population had a competition coefficient of 1.14 while the coefficient for the

Conceptual Breakthroughs in Evolutionary Ecology
ISBN: 978-0-12-816013-8
https://doi.org/10.1016/B978-0-12-816013-8.00055-7

low-density populations was 0.72. That is, the competitive ability of the high-density adapted populations was 58% greater than the low-density populations. A phenotype tightly correlated with competitive ability is the larval feeding rate (Bakker, 1961; Mueller, 1988b). Joshi and Mueller (1988) showed that the high-density adapted populations fed 15% faster than the low-density adapted populations.

Impact: 8

An intuitive but important component of the theory of density-dependent natural selection is the prediction that competitive ability for resources in short supply in a crowded environment will increase by natural selection. The experiments conducted by Mueller (1988a) provide direct evidence for this important theoretical prediction.

References

Bakker, K., 1961. An analysis of factors which determine success in competition for food among larvae in *Drosophila melanogaster*. Arch. Neerl. Zool. 14, 200–281.

Joshi, A., Mueller, L.D., 1988. Evolution of higher feeding rate in *Drosophila* due to density-dependent natural selection. Evolution 42, 1090–1093.

MacArthur, R.H., Wilson, E.O., 1967. The Theory of Island Biogeography. Princeton Univ. Press, Princeton, NJ.

Mueller, L.D., 1988a. Evolution of competitive ability in *Drosophila* due to density-dependent natural selection. Proc. Natl. Acad. Sci. U.S.A. 85, 4383–4386.

Mueller, L.D., 1988b. Density-dependent population growth and natural selection in food limited environments: the *Drosophila* model. Am. Nat. 132, 786–809.

Mueller, L.D., Ayala, F.J., 1981. Trade-off between *r*-selection and *K*-selection in *Drosophila* populations. Proc. Natl. Acad. Sci. U.S.A. 78, 1303–1305.

CHAPTER FIFTY SIX

1990 A predator-prey arms race

The concept

Anti-predator defenses of the newt *Taricha granulosa* are countered by a garter snake that feeds on them. These predator responses appear to be proportional to the level of defense mounted by the newt, consistent with an evolutionary arms race.

The explanation

The skin of the newt *Taricha granulosa* contains the neurotoxin tetrodotoxin (TTX) which is highly toxic to most animals. The garter snake *Thamnophis sirtalis* specializes on amphibians and will eat *T. granulosa* where the two species are sympatric. This suggests that the garter snakes had evolved some tolerance for this toxin. Brodie and Brodie (1990) attempted to study the geographic patterns of TTX resistance. Brodie and Brodie (1990) developed an assay for TTX tolerance that involves injecting snakes with know amounts of TTX and then measuring the snakes sprint speed.

They tested two populations of *T. sirtalis*: one from Oregon that was sympatric with toxic newts and a population in Idaho where newts are not found. In addition, Brodie and Brodie tested a related species of snake, *T. ordinoides* from the same Oregon location, which is not known to feed on toxic newts. Brodie and Brodie found that allopatric populations of *T. sirtalis* and the congeneric species *T. ordinoides* showed similar levels of sensitivity to TTX. However, the sympatric populations of *T. sirtalis* showed substantially elevated resistance to TTX. Brodie and Brodie (1990) conclude that "These facts suggest that TTX resistance is not a property of the genus *Thamnophis* or even of the species *T. sirtalis* at large, but rather has arisen only in populations of *T. sirtalis* that feed on *T. granulosa*".

In a later study Brodie and Brodie (1991) found a population of *T. sirtalis* on Vancouver Island, British Columbia, with TTX resistance between the previously studied Oregon population of *T. sirtalis* and *T. ordinoides*. Such an observation seems to contradict the coevolutionary nature of this TTX resistance. However, a closer examination of Vancouver population of

Conceptual Breakthroughs in Evolutionary Ecology
ISBN: 978-0-12-816013-8
https://doi.org/10.1016/B978-0-12-816013-8.00056-9

newts showed that the toxicity of their skin was at least 1000 times less than the toxicity of the previously studied Oregon population of newts. These observations suggest that the garter snakes evolve a level of tolerance needed to tolerate the TTX but not more.

A third study by Brodie and Brodie (1999) showed that there is a cost to TTX resistance. Brodie and Brodie found that snakes with the greatest resistance to TTX tended to be slower than snakes with lower resistance. Since snakes themselves can be prey, sprint speed is an important phenotype for predator avoidance.

Impact: 7

The work by Brodie and Brodie provide an excellent example of predator-prey coevolution. It also demonstrates the geographic variability of coevolution (Thompson, 1994).

References

Brodie III, E.D., Brodie Jr., E.D., 1990. Tetrodotoxin resistance in garter snakes: an evolutionary response of predators to dangerous prey. Evolution 44, 651–659.

Brodie III, E.D., Brodie Jr., E.D., 1991. Evolutionary response of predators to dangerous prey: reduction of toxicity of newts and resistance of garter snakes in island populations. Evolution 45, 221–224.

Brodie III, E.D., Brodie Jr., E.D., 1999. Costs of exploiting poisonous prey: evolutionary trade-offs in a predator-prey arms race. Evolution 53, 626–631.

Thompson, J.N., 1994. The Coevolutionary Process. University of Chicago Press, Chicago.

CHAPTER FIFTY SEVEN

1990 Reproductive effort and the balance between egg number and egg size

The concept

Barry Sinervo (1990, 1991) developed a cleaver technique for manipulating the energy content of lizard eggs. This allowed him to study basic theories concerning the relationships between egg numbers and egg size.

The explanation

In Chapter 14, we reviewed Lack's Principle which states that birds will adjust their clutch size so as not to lay more than they can successfully raise. Even for animals that do not engage in parental care, a corollary of Lack's Principle applies: due to limited energy animals will either, produce many small eggs or fewer large eggs. The benefit of larger eggs is that they have an increased chance of survival. But how can these conjectures be tested?

It is certainly the case that if a sample of lizards are taken and allowed to lay eggs, those lizards laying fewer, larger eggs will produce larger offspring, and have longer incubation times. It could be that the larger size and longer incubation time are entirely due to the larger egg mass, or it might be that females who produce larger eggs also pass on characters not associated with egg size that affect adult size and incubation time. Sinervo (1990) proposed that yolk size of eggs be artificially reduced, thereby allowing one to manipulate egg size only and study the effects.

Sinervo (1990) studied the relationship between egg size, egg number, and performance traits related to offspring size in the iguanid lizard, *Sceloporus occidentalis*. Life-history theory predicts that a trade-off between these traits will limit their ultimate values but will also change as the environment places different demands on the organism. Such environmental differences were seen in Sinervo's (1990) study; lizards sampled in western states showed a cline in clutch size increasing from 7 in California to 12 in Washington. Accompanying this cline in clutch size was a cline in egg mass, with eggs

Conceptual Breakthroughs in Evolutionary Ecology
ISBN: 978-0-12-816013-8
https://doi.org/10.1016/B978-0-12-816013-8.00057-0

in Washington being smaller than those in California. A similar cline was also seen in California along an altitudinal gradient, with smaller clutches of large eggs being found at lower altitudes.

Also correlated with egg size are several juvenile life-history traits (such as growth rates, hatchling size, incubation time, and hatchling sprint performance). For instance, sprint speed is likely to affect predator avoidance, feeding success, and social dominance. By experimentally reducing egg volume, Sinervo (1990) was able to study the effects of size alone in isolation of other factors. Sinervo found that while egg volume has an effect of growth rates, there exist other genetic and physiological differences between populations that have an independent effect on growth rates. However, differences in sprint speed are largely due to size differences. How can juvenile lizards in Washington survive given their small size and reduced sprint speed? Sinervo suggests that selection for high sprint speed is reduced in Washington relative to California. Thus, in Washington, the benefit of producing more offspring compensates for this reduction in sprint speed.

Sinervo and Licht (1991) next wondered if there is a limit to how large an egg might be that would ultimately control the range of the egg number versus egg size trade-off. Sinervo and Licht studied this by surgically removing yolk from female follicles of the lizard *Uta stansburiana*. A consequence of these experimental removals was the production of fewer, larger eggs. *U. stansburiana* produces on average 4.6 eggs but can produce up to 9 eggs. Using the surgical procedure, Sinervo and Licht were able to produce many females that laid clutches with only 1, 2, or 3 eggs. The eggs from these experimentally manipulated females became larger as the clutch size decreased. As the eggs become larger the frequency of eggs bursting before they hatch increases (Fig. 57.1). Additionally, very large eggs may become bound in the oviduct. In each case, the embryo dies and in the case of bound oviducts the female and any remaining embryos die. These observations suggest there will be strong selection against eggs that are too large. This size limit is most likely set by the size of the lizard's pelvic girdle.

Impact: 8

Through his novel experimental techniques Sinvero has been able to explore and shed light on one of the most interesting life-history trade-offs.

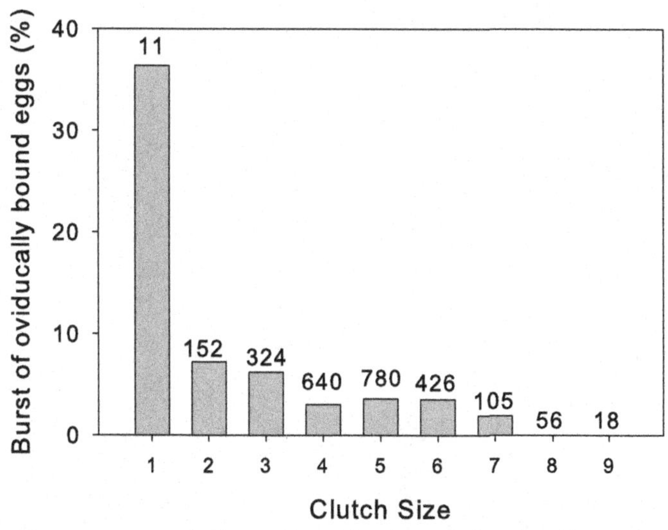

Fig. 57.1 The percentage of all eggs that either burst before hatching or became oviducally bound. Above each bar are the total number of eggs examined in each clutch category.

References

Sinervo, B., 1990. The evolution of maternal investment in lizards: an experimental and comparative analysis of egg size and its effects on offspring performance. Evolution 44, 279–294.

Sinervo, B., Licht, P., 1991. Proximate constraints on the evolution of egg size, number, and total clutch mass in lizards. Science 252, 1300–1302.

CHAPTER FIFTY EIGHT

1991 Experimental evolution of cooperation

The concept

Anderson and May (1982) emphasized the importance of the relationship between parasite virulence and transmission to the ultimate evolution of virulence. Axelrod and Hamilton (1981) suggested that cooperation is more likely to evolve when finding partners is more difficult. Bull et al. (1991) set out to test the evolution of host-parasite fidelity using bacteria and phage structured around the ideas of Anderson, May, Axelrod and Hamilton. They found that when phage are denied opportunities for transmission they evolve reduced virulence for their bacterial hosts.

The explanation

The theory developed by Axelrod and Hamilton (1981) and Anderson and May (1982) suggested that cooperation between species pairs could be enhanced by the manipulation of opportunities for a parasite to find a new host. Bull et al. (1991) decided to study this by looking at bacteria and phage that can form a partner relationship, *E. coli* and phage R208. This phage carries a resistance gene to ampicillin and enters male *E. coli* through the F-pilus. Once infected an *E. coli* cannot be infected by another R208; however, the growth rate of the infected *E. coli* is reduced relative to uninfected cells. The phage reproduce continuously and do not kill their bacterial hosts. Phage progeny may exit the bacteria to infect additional *E. coli* (horizonal transmission) and their genomes can be transmitted to the two daughter cells resulting from *E. coli* reproduction (vertical transmission).

From an initial sample of *E. coli* that had been infected by R208, two selection protocols were followed for about 150 generations. In the high fidelity treatment, infected *E. coli* were allowed to grow in a liquid medium with ampicillin. In this environment any R208 that left the host would be unable to find an uninfected *E. coli* host since they were killed by the ampicillin. There would also be competition among different host/parasites pairs for rapid growth. In the low fidelity treatment, *E. coli* and R208 phage were

Conceptual Breakthroughs in Evolutionary Ecology
ISBN: 978-0-12-816013-8
https://doi.org/10.1016/B978-0-12-816013-8.00058-2

allowed to grow in liquid medium without ampicillin. In this environment R208 that left their host should find many other available hosts to infect.

After this period of adaptation to the two environments, bacteria infected with R208 were isolated from the high and low fidelity cultures and their growth rates measured. The high fidelity line grew roughly 50% faster than the low fidelity line. We can call the improved growth rates of the high fidelity line either increased fidelity, increased cooperation, or reduced virulence. This improvement may have been due to genetic changes in the phage, bacteria or both. Bull et al. (1991) concluded that most of the improvement in the high fidelity lines was due to genetic changes in the phage although a small part of the improvement was also due to genetic changes in the bacteria.

This elegant experiment shows unambiguously that the availability of hosts is a crucial component to the evolution of fidelity. Consequently, this work provides an important building block to our understanding of the evolution of cooperation and the virulence of parasites.

Impact: 10

This experiment contains all the power of modern scientific hypothesis testing to a fundamental problem in species interactions. Consequently, it is a superb example of applying experimental techniques to difficult evolutionary questions.

References

Anderson, R.M., May, R.M., 1982. Coevolution of hosts and parasites. Parasitology 85, 411–426.

Axelrod, R., Hamilton, W.D., 1981. The evolution of cooperation. Science 211, 1390–1396.

Bull, J.J., Molineux, I.J., Rice, W.R., 1991. Selection of benevolence in a host-parasite system. Evolution 45, 875–882.

CHAPTER FIFTY NINE

1994 Experimental test of the role of natural selection in the process of character displacement

The concept

Patterns of divergence consistent with character displacement (Chapter 15) are readily seen in nature. What is more difficult is to document that in sympatry competition between species has fitness consequences that can drive this differentiation. Schluter (1994) carried out experiments with threespine sticklebacks that provided direct evidence of these competitive effects.

The explanation

Threespine sticklebacks (*Gasterosteus aculeatus* complex) inhabit lakes in British Columbia, Canada and are thought to have diversified recently (~13,000 years ago). The geographic patterns of occurrence strongly suggest that character displacement has taken place. In lakes where two species coexist there is a "benthic" species which feeds on benthic invertebrates and has a few short gill rakers and a wide gape while the second "limnetic" species feeds on plankton and has numerous gill rakers and a narrow gape. Lakes with only a single species have sticklebacks with intermediate morphology that utilize both the benthic and plankton habitat.

The ancestral forms of the fish were thought to be the intermediate and limnetic forms. Schluter (1994) wanted to see if he could measure the effects of limnetic species competing with the intermediate species. The intermediate species actually show a range of morphologies from those more similar to benthic species and others more similar to limnetic species. If limnetic species were introduced into a lake with intermediate species it was suspected that individuals from the intermediate species most similar to the limnetic species would suffer the most from competition.

Conceptual Breakthroughs in Evolutionary Ecology
ISBN: 978-0-12-816013-8
https://doi.org/10.1016/B978-0-12-816013-8.00059-4

Schluter utilized two ponds that were each split into two parts. In one-half of each pond he placed only the intermediate species. In the second half of the pond he placed roughly equal numbers of intermediate and limnetic species. The fish competed in these ponds for three months and after that they were all removed and their growth rates measured. Growth rates were used since these are related to components of fitness (such as over-winter size and survival, time of breeding, and fecundity).

Schluter found that the intermediate sticklebacks suffered a reduction in growth rate when competing against the limnetic species and that the magnitude of this reduction was larger the more similar the intermediate morphology was to a limnetic species. Thus, the action of strong competition between the intermediate and limnetic species was demonstrated. This reduction in fitness of intermediate phenotypes similar to the limnetic species would be expected to favor an increase in the frequency of more benthic genotypes in the intermediate species.

Impact: 9

The study by Schluter is an excellent example of applying experimental tests to populations in the field. It also provided some of the strongest evidence for selection as the causative agent of morphological divergence and ecological specialization in the sticklebacks.

Reference

Schluter, D., 1994. Experimental evidence that competition promotes divergence in adaptive radiation. Science 266, 798–801.

CHAPTER SIXTY

1994 The geographic mosaic theory of coevolution

The concept

John Thompson (1994) proposed that a complete understanding of the coevolutionary process required a geographic view. That is, coevolution in local populations was only one part of the story. Due to gene flow, variation in community composition, and other factors, the specific details of coevolution are likely to vary in important details.

The explanation

There is abundant evidence that the outcome of coevolution often varies from one locality to another. Clarke and Sheppard (1971, Chapter 39) point out the many morphs of *Papilio memnon* that resemble different distasteful models. Cuckoo birds are parasites that lay their eggs in the nests of other species who then raise the young of the cuckoo. Parasitized birds will evolve the ability to discriminate between their own eggs and that of the cuckoo. In England meadow pipits and white wagtails are parasitized by cuckoos. However, in Iceland cuckoos are not found and the pipit and wagtail populations there have significantly less discrimination against eggs that are different from their own (Davies and Brooke, 1989).

Thompson (1994) points out that interspecific interactions differ among populations due to differences in the physical environment, local genetic and demographic structure, and the community context. These differences may then lead to coevolutionary responses, or responses by only one species or no responses. If the coevolutionary response results in specialization, it may be to one or multiple species. Since many of the factors that influence coevolution vary over time we expect a continually shifting pattern of coevolution between any two or more species. These different outcomes can also be viewed from the perspective of Wright's shifting balance hypothesis (Wright, 1982). Under this theory, population mean fitness has multiple local peaks and selection may end up climbing any one of them dependent on the initial

Conceptual Breakthroughs in Evolutionary Ecology
ISBN: 978-0-12-816013-8
https://doi.org/10.1016/B978-0-12-816013-8.00060-0

starting conditions. This theory is complementary to Thompson's theory in that it helps us understand the multiple outcomes of coevolution.

Impact: 9

Thompson (1994) does a masterful review of the coevolution literature and then uses it to synthesize a broad theory that makes sense of seemingly contradictory observations. A test of these ideas is reviewed in Chapter 65.

References

Clarke, C.A., Sheppard, P.M., 1971. Further studies on the genetics of the mimetic butterfly *Papilio memnon* L. Phil. Trans. Roy. Soc. Lond. Series B 263, 35–70.

Davies, N.B., Brooke, M. De L., 1989. An experimental study of co-evolution between the cuckoo, *Cuculus canorus*, and its hosts. I. Host eff discrimination. J. Anim. Ecol. 58, 207–224.

Thompson, J.N., 1994. The Coevolutionary Process. University of Chicago Press, Chicago.

Wright, S., 1982. The shifting balance theory of macroevolution. Annu. Rev. Genet. 16, 1–19.

CHAPTER SIXTY ONE

1996 Community evolution due to indirect effects

The concept

Interacting species may have direct effects on each other due to predation or competition or they may have indirect effects that are passed through a chain of species interactions. Miller and Travis (1996) develop predictions about the evolution of traits that affect species interactions based on the impact of these direct and indirect effects.

The explanation

Evolutionary biologists have long acknowledged the important role of close species interactions like those due to competition or predation (see Chapter 20). A species involved in a direct interaction may have a response to the interaction or it may produce an effect that is directed to the other species. For instance, two plant species that grow in close proximity may compete for light. One of those species may *respond* to this competition by becoming, either physiologically or through evolution, tolerant of shade. Alternatively, the same species may, either physiologically or through evolution, grow faster and then have an *effect* on its competitor through shading.

Miller and Travis (1996) look at another dimension where interacting species also feel indirect effects of a third species in their community. One example of such indirect effects is when two prey species are food for the same predator. The prey don't directly interact with each other but, to the extent that one prey can escape the predator, this results in increased predation of the second prey (Holt, 1977). This is an example of a negative indirect effect. A keystone predator results in positive effects on the competing species it controls.

Miller and Travis (1996) then make predictions about the type of evolutionary response that should be expected from a focal species that experiences direct effects from a close associate species and indirect effects from one or more additional species. They predict that when the direct and

Conceptual Breakthroughs in Evolutionary Ecology
ISBN: 978-0-12-816013-8
https://doi.org/10.1016/B978-0-12-816013-8.00061-2

indirect effects on the focal species have the same sign, e.g. they are both negative or both positive, evolution will favor the focal species causing an effect on the close associate in the same direction. However, if the direct and indirect effects on a focal species are opposite in sign, then the focal species should not evolve an effect on the close associate species; rather, it should evolve its own response, e.g. like becoming shade tolerant in the example above.

This line of reasoning suggests that evolution between closely associated pairs of species will to some extent be influenced by the community network of indirect responses and accordingly may vary from one location to another (Thompson, 1988).

Impact: 6

Miller and Travis (1996) paint a complicated picture of the ecological background affecting the evolution of a single species, but one which may be amenable to tests. They suggest some ways in which this theory may be tested while admitting the challenge it presents.

References

Holt, R.D., 1977. Predation, apparent competition, and the structure of prey communities. Theor. Popul. Biol. 12, 197–229.

Miller, T.E., Travis, J., 1996. The evolutionary role of indirect effects in communities. Ecology 77, 1329–1335.

Thompson, J.N., 1988. Variation in interspecific interactions. Annu. Rev. Ecol. Systemat. 19, 65–87.

CHAPTER SIXTY TWO

1996 Tests of adaptive phenotypic plasticity in plants

The concept

The ability of an organism to modify its morphology or physiology in response to environmental variation (phenotypic plasticity) it thought to confer a selective advantage relative to individuals unable to make such changes. Dudley and Schmitt (1996) show that a plant's phenotypic response to crowding indeed increases its fitness at the density that elicited the response.

The explanation

Many plants, including *Impatiens capensis*, experience reduced light as a result of competition from other plants especially under crowded conditions. A response to this shading is elongation of the plant's branches presumably in search of more direct sunlight. To show that this response is adaptive requires a demonstration that fitness of the elongated plant is higher in a crowded environment than an unelongated plant in the same environment.

The elongation response of *I. capensis* can be stimulated by manipulating the ratio of red to far-red light the plant receives. Shading by plant competitors alters light in the same way. Dudley and Schmitt (1996) grew plants in crowded conditions and created two treatments. In the elongation treatment, ambient light was at a normal red to far-red ratio which was then reduced by the shading of competing plants which elicited the elongation response. In the suppressed treatment the red to far-red ratio was elevated so that the elongation response was suppressed. Plants from both treatments were then raised in replicated plots under high and low plant densities.

For each plant Dudley and Schmitt (1996) measured lifetime fitness as the total number of reproductive structures, fruits, flowers and pedicels, over an individual plant's lifetime. Dudley and Schmitt also estimated selection differentials and found that at high density there was selection for increased plant height and leaf length. At low density there was selection for increased leaf length but decreased height. The most important observation was on lifetime

Conceptual Breakthroughs in Evolutionary Ecology
ISBN: 978-0-12-816013-8
https://doi.org/10.1016/B978-0-12-816013-8.00062-4

fitness. At low density the suppressed plants had higher fitness than the elongated plants. However, at high density the opposite was seen: elongated plants had higher fitness than suppressed plants. These last observations show that the phenotypic response of individual *I. capensis* to crowding increases their fitness. This is strong evidence for the adaptive basis of phenotypic plasticity.

Impact: 8

This work utilized individuals from natural populations of *I. capensis* and a straightforward experimental design that provided unambiguous evidence that this particular phenotypic response to density is adaptive. While the idea that phenotypic plasticity is adaptive is not new (Bradshaw, 1965) Dudley and Schmitt provide some of the best evidence.

References

Bradshaw, A.D., 1965. Evolutionary significance of phenotypic plasticity in plants. Adv. Genet. 13, 115–155.

Dudley, S.A., Schmitt, J., 1996. Testing the adaptive plasticity hypothesis: density-dependent selection on manipulated stem length in *Impatiens capensis*. Am. Nat. 147, 445–465.

CHAPTER SIXTY THREE

1997 Evolution of species' range – theory

The concept

The geographical range of a species is often thought to be determined by resource requirements, physical habitat limitations, or competitors. However, species can also adapt to these limiting factors and expand their range. Kirkpatrick and Barton (1997) point out that migration from the center of a species' distribution may overwhelm populations on the periphery and prevent adaptation. Kirkpatrick and Barton's theory explores the conditions under which this may occur.

The explanation

Plants and animals often have characteristic geographical ranges that demarcate where they can live. Sometimes the boundary of a range is determined because mortality is 100% beyond the boundary. Thus, a terrestrial plant's territory will end at the coastline because the plant cannot live in saltwater. In other cases, the factors determining the boundary are less clear. This raises the question: why doesn't the range expand via adaptation into the inhospitable area? In chapter 35 we have seen that mine tailings can initially create a sharp boundary for grass species. Over time, however, individuals with adaptations that permit them to survive in this new environment arise and expand the geographical boundary of the plant.

Kirkpatrick and Barton (1997) raise the possibility that gene flow from the center of the species distribution will prevent such adaption of the peripheral populations since the central populations are maladapted to the environment on the periphery. Kirkpatrick and Barton model the process of dispersal, adaptation, and population regulation. The mathematics of the model are complicated so what follows is a brief verbal summary of the model and its results.

The geographical range is modeled as a single linear dimension. Following birth, individuals disperse at random although the mean distance of dispersal is

Conceptual Breakthroughs in Evolutionary Ecology
ISBN: 978-0-12-816013-8
https://doi.org/10.1016/B978-0-12-816013-8.00063-6

controlled by a model parameter. At each location in the range there is an optimal value for a quantitative genetic trait that is under selection. That optimum can change slowly with distance or it can change rapidly. This changing optimum is called the environmental gradient. The steepness of the environmental gradient is measured by a parameter *b*. Population size is a function of the average fitness of phenotypes at each location. The further a phenotype is from the optimum, the lower the phenotype's fitness.

The model shows that if the environmental gradient changes very slowly with distance, near-perfect adaptation is possible at all locations and the range expands forever. However, if the gradient becomes very steep, then range expansion can be halted by the inability of the populations at the edge of the range to adapt further. Additionally, with a steep gradient there will also develop a cline in the phenotype under selection — that is, the mean phenotype will linearly decrease (or increase) as it moves away from the center of the range.

It is worth noting that in Chapter 35 the expansion of grasses into soils with heavy metals occurred over very short distances despite documented gene flow. However, the plants in the contaminated soils showed an increased frequency of selfing. Selfing would naturally be expected to reduce gene flow and by this model facilitate additional adaptation to heavy metals.

Impact: 8

Kirkpatrick and Barton give support for the notion that dispersal can limit species' range expansion. By creating a formal model of the process they highlight the important aspects of selection and population dynamics that may control this process, and be subject to empirical testing.

Reference

Kirkpatrick, M., Barton, N.H., 1997. Evolution of a species' range. Am. Nat. 150, 1–23.

CHAPTER SIXTY FOUR

2000 Evolution of population stability: experiments

The concept

The theory described in Chapter 47 did not result in a simple, robust understanding of the evolution of population stability. What was needed were some straightforward experimental tests. Mueller and colleagues (Mueller et al., 2000; Mueller and Joshi, 2000) created large populations that exhibited unstable population dynamics. Over 68 generations these populations showed evidence of phenotypic change as they adapted to these environments, but population stability remained unchanged suggesting population stability does not directly respond to selection. Prasad et al. (2003) found that selection for rapid development reduces female fecundity as a correlated trait and this evolution also produces more stable population dynamics. Hence with the appropriate trade-offs the evolution of enhanced stability is possible.

The explanation

One of the earliest and most detailed studies of a population with unstable dynamics is due to Nicholson (1954, 1957) using laboratory populations of blowflies. Nicholson's data typically consisted of long (400–700 days) observations of adult and egg numbers in unreplicated populations. Stokes et al. (1988) analyzed one of these long experiments by dividing the data into two parts: population size before day 500 and those after day 500. This division was created since the magnitude of fluctuations appeared to diminish after day 500. Stokes et al. (1988) concluded that in fact the dynamics of the population had become more stable over time. They attribute this to a reduction in the maximum fecundity of females and the consequent stabilizing effects of such a reduction. However, another interpretation of the changes over time is that due to the many bottlenecks in population size the adults had become inbred and that such inbreeding reduced female fecundity. In other words, selection was not the primary driving force behind the increased stability of these blowfly populations.

Conceptual Breakthroughs in Evolutionary Ecology
ISBN: 978-0-12-816013-8
https://doi.org/10.1016/B978-0-12-816013-8.00064-8

A detailed life-historical model of *Drosophila* population dynamics suggested that stability in a discrete-generation life cycle depended on the relative amounts of food provided to the adults and larvae (Mueller, 1988). Unstable dynamics occurred when larvae were provided low levels of food and adults were provided high levels. The opposite set of conditions produced stable dynamics (Mueller and Huynh, 1994).

To properly test whether natural selection may favor increased stability rather than stability being a byproduct of inbreeding Mueller et al. (2000) set up 10 replicate populations of *Drosophila melanogaster* with unstable cycling populations and an equal number with stable population dynamics. However, these populations rarely had adult populations below 1000 individuals, hence eliminating the chance of inbreeding. While there was ample evidence of phenotypic change in these populations due to genetically based changes during adaptation, there was not an increase in the stability of the population dynamics. While it was possible that 68 generations (1428 days) was not sufficient time to observe a slow increase in stability (given that other traits were responding to selection) suggests that adaptation in *Drosophila* to environments that cause unstable dynamics does not result in increased population stability.

In an unrelated study, Prasad et al. (2003) had studied the phenotypic responses to selection for rapid egg-to-adult development in *D. melanogaster*. They found that the rapidly developing populations suffered a 35% reduction in fecundity relative to controls. When the populations selected for rapid development and the control were then placed in environments known to cause unstable dynamics, the rapidly developing lines showed enhanced stability relative to controls. Mueller and Ayala (1981) concluded from simple theoretical models that populations exhibiting unstable dynamics could evolve stable dynamics if female fecundity could be reduced. Normally, natural selection would not favor the reduction in female fecundity, even at high population density. However, when there is strong selection for other life-history traits, like development time, fitness might on balance increase even with a reduction in fecundity. In this setting, stable population dynamics winds up being a correlated trait which changes in response to changes in development time and fecundity.

Can natural selection lead to reductions in fecundity in nature? *Drosophila sechellia*, a recent and close relative to *Drosophila simulans*, has significantly lower fecundity than *D. simulans*. Mueller and Bitner (2015) argue that this reduction is a direct result of selection for more rapid development of *D. sechellia* larvae as an adaptation to their specialized food resource (morinda fruit). In this case *D. sechellia* accomplishes faster development by laying eggs that hatch in 2 h rather than 24 h.

Impact: 8

This study (Prasad et al., 2003) was important for competing the circle with previous theoretical and empirical work on this problem. It shows that consistent with some previous theory, natural selection can increase population stability but only if there is sufficiently strong selection for reduced fecundity. Studying the evolution of a population level property like stability is only possible in a model system. Thus, this work also highlights the importance of utilizing such experimental systems in evolutionary ecology research.

References

Mueller, L.D., 1988. Density-dependent population growth and natural selection in food limited environments: the *Drosophila* model. Am. Nat. 132, 786–809.

Mueller, L.D., Ayala, F.J., 1981. Dynamics of single species population growth: stability or chaos? Ecology 62, 1148–1154.

Mueller, L.D., Bitner, K., 2015. The evolution of ovoviviparity in a temporally varying environment. Am. Nat. 186, 708–715.

Mueller, L.D., Huynh, P.T., 1994. Ecological determinants of stability in model populations. Ecology 75, 430–437.

Mueller, L.D., Joshi, A., 2000. Stability in Model Populations. Monographs in Population Biology. Princeton University Press, Princeton, N.J.

Mueller, L.D., Joshi, A., Borash, D.J., 2000. Does population stability evolve? Ecology 81, 1273–1285.

Nicholson, A.J., 1954. An outline of the dynamics of animal populations. Aust. J. Zool. 2, 9–65.

Nicholson, A.J., 1957. The self adjustment of populations to change. Cold Spring Harbor Symp. Quant. Biol. 22, 153–173.

Prasad, N.G., Dey, S., Shakarad, M., Joshi, A., 2003. The evolution of population stability as a by-product of life-history evolution. Proc. R. Soc. Lond. B 270, S84–S86.

Stokes, T.K., Gurney, W.S.C., Nisbet, R.M., Blythe, S.P., 1988. Parameter evolution in a laboratory insect population. Theor. Popul. Biol. 34, 248–265.

CHAPTER SIXTY FIVE

2004 Coevolution over space and time

The concept

Natural environments vary over time and space. Coevolution of species interactions will almost certainly be affected by this variation. Forde et al. (2004) show that these ecological complexities can be studied in a systematic way and that the dynamics of the coevolutionary responses are predictable.

The explanation

The geographic mosaic theory of coevolution (Chapter 60, Thompson, 1994) suggests that the dynamics of coevolution will be affected by geographic differences in resources, community composition, and migration. Thus, finding pairs of species at different stages of coevolution is not evidence against coevolution but the expected pattern. To study this, Forde et al. (2004) created populations of *Escherichia coli* and T7 phage and followed their coevolution. T7 infect *E. coli*, replicate and then kill the host as they emerge. *E. coli* and T7 each go through a progression of changes as they coevolve.

Let the starting genotypes of *E. coli* and T7, be designated B_0 and $T7_0$ respectively. The natural progression of coevolution is for host resistance to $T7_0$ to develop in a mutant genotype designated B_1. Somewhat later the phage mutant, $T7_1$, called a host range mutant will appear. $T7_1$ can infect either B_0 or B_1. Lastly, B_1 can mutate to become resistant to $T7_1$. This mutant is called B_2.

Forde et al. (2004) studied the coevolution of phage and host by creating three different environments that varied in productivity – high, intermediate and low. It has previously been documented that higher productivity environments increase the rate of appearance of phage resistant bacteria (Bohannan and Lenski, 1997). On top of this structure were two different migration regimes: no migration (closed) and a stepping stone pattern of high productivity to intermediate productivity to low productivity (open).

Conceptual Breakthroughs in Evolutionary Ecology
ISBN: 978-0-12-816013-8
https://doi.org/10.1016/B978-0-12-816013-8.00065-X

Adaptation in each environment was measured by quantifying phage infectivity in samples from their evolved environment (sympatry) to infectivity in an independent environment (allopatry) with the same productivity. Levels of adaptation were quantified at 9, 13, and 19 days after the start of the experiment.

Adaptations appeared first in high productivity environments. $T7_1$ appeared after 5 days in the high productivity environment, 7 days in the intermediate and 11 days in the low productivity environment. Levels of adaptation were consistently higher in landscapes with migration. In an open system there was a cascade of adaptation that was time-lagged and followed the direction of migration. Forde et al. conclude that "The results demonstrate that even in a relatively simple, controlled experimental setting, coevolutionary interactions are highly dynamic when embedded within a geographic mosaic".

Impact: 7

A major critique of using laboratory evolution to study ecological questions is that these laboratory environments are overly simple. Forde et al. have incorporated complexity into the laboratory system and demonstrated that evolution is profoundly impacted by this ecological complexity.

References

Bohannan, B.J.M., Lenski, R.E., 1997. Effect of resource enrichment on a chemostat community of bacteria and bacteriophage. Ecology 78, 2303–2315.

Forde, S.E., Thompson, J.N., Bohannan, B.J.M., 2004. Adaptation varies through space and time in a coevolving host–parasitoid interaction. Nature 431, 841–844.

Thompson, J.N., 1994. The Coevolutionary Process. University of Chicago Press, Chicago.

Epilogue

No story of evolution is complete without an ecological context, but the environment is complex and varied, so much remains to be done in the field of evolutionary ecology. Certainly, advances in technologies like genomics will aide in developing our understanding of the evolutionary process. However, technologies without good questions are of little help. Evolutionary ecology has benefited from a mature interaction between theory and experiments. This will hopefully continue.

Epilogue

Appendix

Evolutionary ecology research involves a combination of techniques shared by other disciplines but with some historical preferences and biases. Like population genetics, there is a well-developed body of theory in evolutionary ecology. Evolutionary ecology also utilizes field research common to ecology. In addition, there is a substantial contribution to our understanding of evolutionary ecology from experimental research, the backbone of many disciplines in biology, chemistry, and physics.

In this appendix I will review issues that evolutionary ecology has grappled with in the development of theoretical and experimental research. Many important advances in evolutionary ecology utilize measures of genetic change or genetic variation. I will also review the most commonly used genetic techniques in evolutionary ecology that have changed substantially over the years.

THEORY

Theory in evolutionary ecology borrows many of the techniques common to population genetics. However, many assumptions common to population genetic models, constant environments, no age-structure, and an absence of any consequential impact of other species are inimical to evolutionary ecology. Certainly, an important component of theory in evolutionary ecology is the exploration of models which add important ecological details often ignored in population genetic theory. Next, I review some of the basic issues faced by evolutionary ecologists when developing theory.

Life-cycle

The timing and pattern of reproduction, referred to as the life history of an organism, falls into two broad categories: semelparity and iteroparity. A semelparous organism reproduces once after development and sexual maturation are completed. An iteroparous organism may reproduce multiple times during its adult lifespan. Some populations of semelparous organisms have synchronized timing of reproduction. Examples include annual plants and salmon. Certainly, in the laboratory organisms which are not semelparous in nature can be forced to reproduce synchronously and only once. In other cases, the semelparous organisms may reproduce more or less continuously, like bacteria or protozoans.

In any genetic model the genetic state of the population is followed over time. The genetic state in a single-locus model with two alleles would be just the frequency of one of those alleles. If the genetic model involves multiple loci and alleles then there will be a vector of haplotype frequencies that will be followed. For a quantitative genetic model the genetic state will be a vector of mean trait values and an additive variance-covariance matrix. In any case we summarize the genetic state at time-t with the symbols $\mathbf{g}(t)$.

In the case of a semelparous organisms that reproduces synchronously, as in Fig. 1A, the genetic state at time-$(t + 1)$, will be some function of the

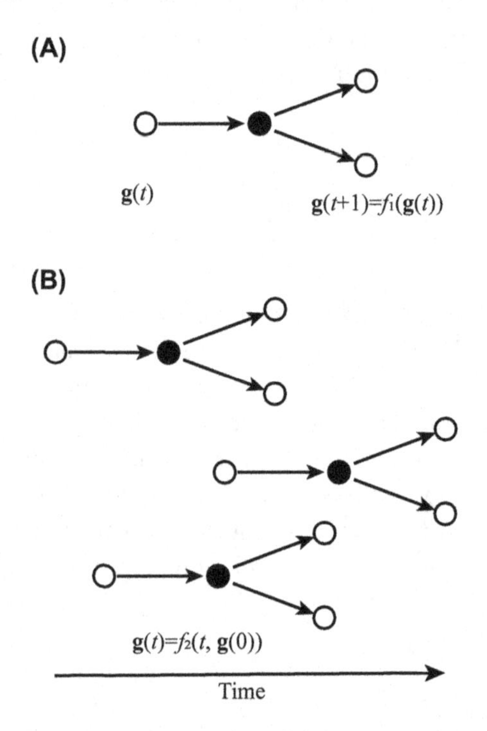

Figure 1 Common life cycles used in models of evolutionary ecology. The *open circles* are progeny and the *filled circles* are reproductive adults. Although each adult has only two offspring in this illustration, in general the number will vary between individuals. (A) A discrete generation life cycle where t indicates the generation number. The genetic state vector is $\mathbf{g}(t)$. The value of this vector depends only on the genetic values in the previous generation. (B) Continuous reproduction. In this case individuals are reproducing asynchronously, and the genetic state of the population depends on the starting state and the amount of time that has passed.

genetic state at time-t, or $\mathbf{g}(t+1) = f_1[\mathbf{g}(t)]$ (Fig. 1A). Examples of models utilizing this life cycle can be found in Chapters 30, 31, and 36. The function, f_1, contains the technical details of how you move from one time point to the next and information like fitnesses. For models of the type outlined in Fig. 1A, time is measured in generations.

For semelparous organisms that reproduce continuously, time will be measured in standard units, e.g. minutes, months, years (Fig. 1B). Models of genetic change for this life cycle require knowledge the starting genetic state, $\mathbf{g}(0)$, and the time that has passed, $\mathbf{g}(t) = f_2[t,\mathbf{g}(0)]$. Examples of models using continuous time are in Chapters 41 and 45.

Iteroparous organisms may have synchronized breeding episodes or seasons, like birds in temperate environments, or they may also reproduce more or less continuously (Fig. 2). Even if reproduction is continuous, time may be broken into discrete intervals, called age-classes, which are then used to summarize the survival and fertility of individuals whose age falls within an age-class. The genetic state of the population in an iteroparous organisms must then include newborns plus all the surviving adults (Fig. 2). If adults can survive for up to d time intervals then a complete description of the genetic state of the population will need to go back d time units into the past (Fig. 2).

Needless to say, this makes the analysis of age-structured populations much more complicated than those without age-structure and in part

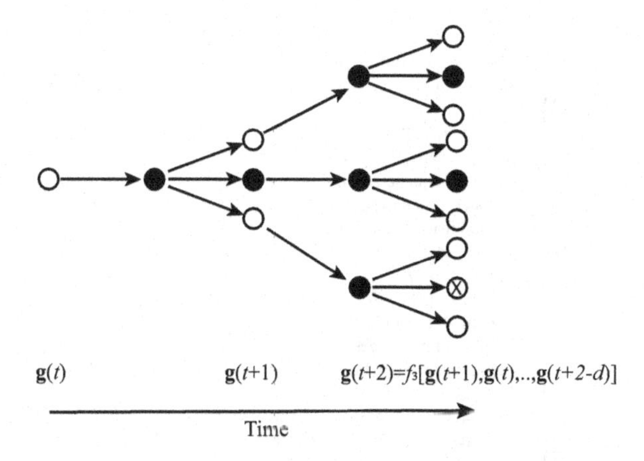

Figure 2 The life-cycle of an age-structured population. The symbols are the same as in Fig. 1 except the "X" in a circle indicates an adult who has died. The maximum lifespan of this organism is d time units. The genetic state of the populations depends on the previous d genetic states.

explains the focus on models without age-structure. A continuous time age-structure model was reviewed in Chapter 4.

Genetic model

The creation of a model of evolution requires some underlying assumption about the genes controlling the traits that are under selection. A summary of the range of genetic models typically encountered in given in Fig. 3.

Explicit genetic models assuming 1, 2, or 3 loci draw from our detailed understanding of genetics (Fig. 3A–C). Single locus models are common and generally easier to analyze (Fig. 3A). If there are only two alleles at the single locus then only one allele frequency has to be modelled. Single locus models are used for the theory reviewed in Chapters 29, 34, 39, and 47. With two loci, each with two alleles, there are now four chromosome frequencies that require three systems of nonlinear-equations to analyze (since one chromosome frequency must always equal 1 minus the sum of the other three, Fig. 3B). Fitness values may be required for up to 10 genotypes if cis and trans double heterozygotes have different fitness values. The addition of recombination also complicates the analysis. A two-locus model is used in Chapter 39.

Genetic Models

(A) Single locus
alleles: $A_1, A_2, \ldots, A_k$
allele frequencies: $p_1, p_2, \ldots, p_k$
number of genotypes: $k(k+1)/2$

(B) Two locus
alleles: $A, a; B, b$
gametes: AB, Ab, aB, ab
gamete frequencies: x_1, x_2, x_3, x_4
number of genotypes: 10
recombination fraction: r

(C) Three locus
alleles: $A, a; B, b; C, c$
gametes: $ABC, ABc, AbC, Abc, aBC, aBc, abC, abc$
gamete frequencies: $x_1, \ldots, x_8$
number of genotypes: 31
recombination fractions: r_1 (AB); r_2 (BC); $r_1 + r_2 - 2r_1r_2$ (AC)

(D) Many loci
Phenotypic variance-covariance matrix: V_p
Additive variance-covariance matrix: V_a

(E) No genetics
Phenotype with hypothesized fitness relationships

Figure 3 The range of genetic models typically utilized in evolutionary ecology research. The relative advantages of each is described in the text.

I don't know of any evolutionary ecology research which has explicitly used a three-locus model although these have been studied by population geneticists (Feldman et al., 1974). These models will be very complicated. Even with just two alleles at each locus there will be 7 non-linear recursion equations to analyze with up to 31 different fitness values unless simplifying assumptions are made (Fig. 3C).

What value is added by looking at two or more loci? With linked loci it is possible to have multiple simultaneous stable equilibria (Karlin, 1975). The existence of multiple equilibria are the basis for Wright's shifting balance theory of evolution (Wright, 1982). In Chapter 30 an important evolutionary question was whether evolution would favor a decrease in recombination between multiple loci and thus lead to a supergene complex controlling mimicry. However, for many traits of interest to evolutionary ecologists it is assumed that many loci affect the trait and thus theoreticians have moved from genetically explicit models to quantitative genetic models.

The quantitative genetic approach, used in Chapter 51, rather than following individual allele frequencies, the genetics of the phenotypes of interest are summarized by the additive genetic variance-covariance matrix (Fig. 3D). This information can be used to predict phenotypic changes in traits directly under selection as well as the response of traits that are correlated to the trait of interest. The additive variance will diminish over time unless it is replaced by new mutations (which is unlikely if selection is very strong).

A final modelling approach is not to use any underlying genetic model to study evolution (Fig. 3E). Instead, the evolution of one or more phenotypes is studied through their hypothesized relationship to fitness. Typically, an evolutionary solution is sought that will maximize fitness. This approach was used in Chapter 26. As discussed in Chapter 26, this approach is motivated by Fisher's Fundamental Theorem of Natural Selection which suggests that selection will carry a population to an equilibrium where fitness is maximized. However, as pointed out in Chapter 26 there are population genetic counterexamples to the fitness maximization prediction of Fisher.

There are several defenses of the optimization approach to studying evolution (Oster and Wilson, 1978; Stephens and Krebs, 1986, chpt. 10). Certainly, an advantage of optimality models is the production of precise predictions. Stephens and Krebs (1986) note that optimization is often criticized for adding more complicating features in an ad hoc fashion when the predictions of an optimality model are not met. They correctly point out that other branches of science will formulate new hypotheses or models when faced with empirical evidence resulting in the rejection of their current idea. However, a major issue with any non-genetic, optimality model can be that it fails to predict the observed outcomes of evolution because evolution has not resulted in an optimal design — no degree of fiddling with the model details or complexity can make a non-optimal solution appear optimal except through faulty reasoning.

The classic example is overdominance. A heterozygote with the highest fitness will cause a population to equilibrate at an equilibrium in which both of the inferior homozygotes are continually produced. There can be no population consisting of 100% heterozygotes causing the population to show less than perfect adaptation.

General vs. specific models

Certainly, a goal of any theory is to be subjected to empirical tests. There can often be problems in trying to test theories which assume simple life-cycles, like that shown in Fig. 1A, with organisms having more complicated life-cycles (Mueller, 1997). Consider simple models of population growth like the logistic equation (Chapter 3). This model predicts the population size from one generation to the next. However, for many organisms the census stage for the logistic model is unstated. As an example, when we take fruit flies into the laboratory and enforce a fully discrete life cycle with only one opportunity for adults to reproduce, there are still pre-adult life stages including eggs, larvae, and pupae. With this simple life-cycle, density dependent control of population growth may happen in several different ways. Density-dependence may affect survival of larvae and/or pupae, it may affect survival of adults or it may affect adult fertility. As it turns out, in fruit flies adult fertility may be affected by the current adult density or the larvae density at which the adult was raised (Mueller, 1985).

Consider a simple life cycle with two possible census stages; eggs and adults (Fig. 4A). The survival of eggs to adults may be affected by the density of eggs or it may be affected by genetic differences among the pre-adult stages. Likewise, the fertility of the adults may also be affected by the density of adults or genetic differences between adults. Prout (1980) looked at two different cases. In the first, survival was density-dependent and fertility was density-independent but showed genetically based variation (Fig. 4B). Secondly, the survival may be a function of genetic differences but density-independent and fertility may be density-dependent (Fig. 4C).

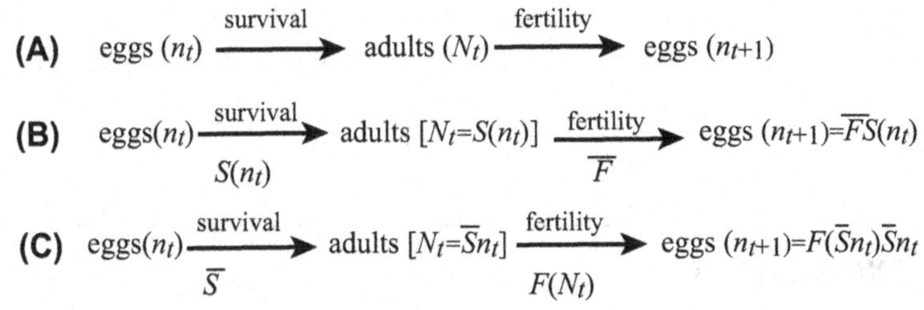

$$\text{(A)}\quad \text{eggs } (n_t) \xrightarrow{\text{survival}} \text{adults } (N_t) \xrightarrow{\text{fertility}} \text{eggs } (n_{t+1})$$

$$\text{(B)}\quad \text{eggs}(n_t) \xrightarrow[S(n_t)]{\text{survival}} \text{adults } [N_t{=}S(n_t)] \xrightarrow[\overline{F}]{\text{fertility}} \text{eggs } (n_{t+1}){=}\overline{F}S(n_t)$$

$$\text{(C)}\quad \text{eggs}(n_t) \xrightarrow[\overline{S}]{\text{survival}} \text{adults } [N_t{=}\overline{S}n_t] \xrightarrow[F(N_t)]{\text{fertility}} \text{eggs } (n_{t+1}){=}F(\overline{S}n_t)\overline{S}n_t$$

Figure 4 Life-cycles in which population size is affected by density-dependence and genetic variation. (A) The number of eggs at generation t is n_t, and the number of adults is N_t. (B) Egg-to-adult survival may be density-dependent and survival controlled by the function $S(n_t)$. Fertility is density-independent but varies among genotypes. The population mean fertility, $\overline{F}$, determines the average number of eggs produced per individual in the population. (C) Egg-to-adult survival is density independent but varies among genotypes. The average survival in the population is given by the mean survival, $\overline{S}$, of all genotypes. Fertility is density-dependent and controlled by the function $F(N_t)$.

Prout (1980) showed that outcome of selection is to maximize the equilibrium number of eggs in the first case (Fig. 4B) and maximize the equilibrium number of adults in the second case (Fig. 4C). However, depending on the form of the density-dependent equations, adult population size at equilibrium may not be maximized in the first case (Fig. 4B) and the equilibrium number of eggs may not be maximized in the second case (Fig. 4C). Thus, results by Roughgarden (1976; Chapter 40) concerning population size maximization are sensitive to the details of the life-cycle and how density-dependence acts.

Some problems require even more specific models when testing theory on specific organisms. For instance, intra-specific competition may have many mechanisms (Keddy, 2001). The specific mechanisms used for securing resources often vary from species to species and therefore the way such populations evolve in response to limiting resources are expected to vary. In Chapter 55, experiments testing the evolution of competitive ability in *Drosophila melanogaster* are described. The measurements of competitive ability and the predictions that it will increase due to selection in a food limited environment were based on a model which specifically included the mechanisms of competition in *Drosophila* (Mueller, 1988).

Another example of taking into account organism-specific information involved modelling the population dynamics of *Tribolium* (Desharnais and Costantino, 1982; Dennis and Costantino, 1988; Peters et al., 1989). In this case, the detailed population dynamic model was ultimately used to predict population stability and compare to experimental populations (Dennis et al., 1995).

EXPERIMENTS

Included in the repertoire of the evolutionary ecologists are laboratory experiments as exemplified by the research in Chapters 8, 46, 50, 52, 58, 64, and 65. Research in natural environments include experiments, see Chapters 31, 48, 59, and 62, and observations and measurements, as in Chapters 1, 2, 9, 10, 12, 27, 53, and 56. Each has different strengths and weaknesses, although to be used properly each method should follow some basic rules. I review some of these issues next.

Laboratory vs. field experiments

I have previously outlined some of these issues (Mueller and Joshi, 2000, chpt. 1) but will add to that discussion here. Most disciplines in biological sciences utilize experimental techniques carried out in controlled laboratory settings. Molecular and developments biologists would never tolerate doing experiments under uncontrolled and randomly varying temperatures as we might find in a natural environment. The obvious reason is that such a procedure will introduce uncertainty likely to affect the experimental outcomes and obscure the impact of the true variables under study.

Research in evolutionary ecology can benefit from the same ability to control all factors except those under study. Carefully designed experiments can be used to test theories in evolutionary ecology which by their very nature are often simple and make many assumptions about the environmental variables that matter. However, the counter to this approach is that these laboratory environments are overly simplistic and can't possibly capture the complexity of nature.

While the previous objection is technically correct, there is a corollary of that objection that is applicable to field experiments. For instance, if I study some evolutionary process in the Sonoran Desert in 2019 what is to say that this environment is the same as it was, in all respects, to the Sonoran Desert in 1990? Is the environment of the Sonoran Desert in 2019 the same as the environment in the Atacama Desert in 2019? Thus, if the "natural" conditions of a field experiment are specific to a precise time and location, they lack the ability to help us develop general principles. While I think I have overstated the objections to field research I do want to emphasize that there is no such thing as "the natural environment", rather there are effectively an infinite number of natural environments which can never be completely characterized.

As an example of when field research is valuable and when it is questionable, first consider the study of Darwin's finches by Peter Grant and colleagues (Chapter 53). This research brings together natural history, ecology and genetics to develop a compelling story of how a changing environment and natural selection have modified the phenotypes of finches on the Galápagos Island of Daphne Major. Although the evolution seen in the 1970's on Daphne Major may never happen again and there may be no other populations of birds that experience exactly this type of evolution, the understanding gained from this detailed study is nevertheless of great value.

Age-specific life history patterns have also been studied in the laboratory (Chapter 50) and in the field (Chapter 48). Recently (Jones et al., 2014), some scientists have studied age-specific patterns of mortality from natural populations as a means of understanding aging. Of course, aging is generally defined as the increase in mortality with increasing age that is due to the deterioration of an organism's physiological health. Mortality in nature may certainly be due to aging, but is also greatly affected by predators, parasites, and vagaries of the environment (Gewin, 2013) (factors that cannot be controlled in the field). In addition, these environmental sources of mortality may change over time and thus confound age-related sources of mortality. Taking samples of various aged individuals at a single point in time is also not a cure for this effect. It has been known for some time (Pearl, 1927) that, for instance, past episodes of crowding may affect survival later in life.

Laboratory experiments

The most successful use of laboratory experiments in evolutionary ecology is to use the laboratory to create environments to which populations then adapt (see for instance Chapter 46). These experiments are properly called "laboratory natural selection". Some researchers mistakenly call them artificial selection since the environment is artificial. Artificial selection is usually reserved for a process in which the experimenter makes measurements of some phenotype and based on these measurements chooses the parents to produce the next generation. This is the technique most often used for selection of important agricultural animals or crops. Simple as these experiments sound there are some important guidelines that must be followed to avoid confounding the interpretation of results (Rose et al., 1996).

Evolution depends on genetic variation. In the case of haploid asexual populations, that genetic variation will be generated by mutations. But for sexually reproducing populations, the founding populations will be the source of genetic variation upon which selection will act. For this reason, a large sample is needed to start populations. For instance, Baldwin-Brown et al. (2014) suggest a starting population of at least 500 haplotypes (which for a sexual, diploid population would be 125 males and 125 females). Before any experimental treatment is initiated, the samples from the wild populations should be allowed time to adapt to the new laboratory environment. If adaptation to the new environment and the experimental selection are attempted at the same time, then the population may move to a different equilibrium than it would have reached had the population first adapted to the lab environment. The complexities of fitness landscapes are discussed in Chapter 47. Even with two-locus models the appearance of multiple equilibria with different domains of attraction are seen (Karlin, 1975). These considerations suggest the fitness landscape may be very different when a population adapts to both the lab environment and, say, high density vs. just high density.

The next important issue is the size of each individual population. If the population size is too small, then the progress of selection may be thwarted by drift, and fitness gains may be obscured by inbreeding and the inevitable fitness decline. In finite populations, alleles favored by selection may still be lost, depending on their initial allele frequency, strength of selection, and effective population size (Mueller et al., 2013). Thus, if the population size is small then adaptation in the laboratory can only occur if genetic variants have large effects on fitness. In other words, small populations will limit the potential of laboratory populations to effectively adapt to novel conditions.

Likewise, inbreeding can lead to an increase in homozygosity (or a decline in heterozygosity). If recessive deleterious alleles are made homozygous, they are liable to have large effects on many fitness-related traits. In *Drosophila*, for instance, the impact of inbreeding has been well documented on egg-to-adult survival (Dobzhansky et al., 1963) as well as fertility in both males (Britnacher, 1981) and females (Marinkovic, 1967). Heterozygosity for neutral genetic markers is reduced at a rate of $\left[1 - \frac{1}{2N}\right]^t$ if N is the effective population size and t is the number of generations. Thus, if

$N = 1000$, heterozygosity at a neutral locus will be reduced by about 5% in 100 generations and 40% in 1000 generations. Thus, another important component of experimental design is how long the selection experiment is expected to last.

Population size can have detectable effects. In the experiment reported by Mueller (1987), one set of populations was kept at a breeding population size of 50. After more than 120 generations of selection, these populations would be expected to have lost about 70% of their original heterozygosity at neutral loci. Indeed Mueller (1987) found that there was evidence of a reduction in female fecundity as a result of this sort of inbreeding. In sexually reproducing diploid organisms, a population size of about 1000 is now seen as a reasonable level to use in laboratory experimental evolution studies (Baldwin-Brown et al., 2014).

A great strength of experimental evolution is the ability to replicate entire populations. Whole populations even when drawn from the same source will always exhibit genetic differences due to sampling effects and drift over time. By maintaining replicate populations, it is possible to estimate the magnitude of these non-adaptive genetic differences. When phenotypes are the primary target of selection, 3–5 replicates of experimental and control populations are reasonable (Rose et al., 1996). However, now that genomic data on whole populations can be collected, yielding massive amounts of data, the assessment of sample size has been re-examined (Baldwin-Brown et al., 2014; Mueller et al., 2018). These studies suggest that to be reasonably certain of detecting genetic differences, at least 50–60 total populations (half controls and half experimentals) should be used.

Lastly, the creation of good controls is essential. The control population should be identical to the experimental except for those elements of the environment that one wants to vary. This is not always simple to carry out. For example, if one wants to explore the effects of adult density on the evolution of life history traits, one cannot create a control without an adult density. You can vary the densities, say low and high, but even here what constitutes low and high has to be carefully considered. For instance, Mueller et al. (1993) found that a "high" larval density of 500 eggs per 6-dram vial did not result in measurable phenotypic changes after 12 generations. Increasing the density to 1000 per 6-dram vial resulted in the phenotypic differentiation that had been seen in previous experiments.

Occasionally, very specific critiques of laboratory selection experiments have been raised. Since some of these come from well-regarded scientists, we review two of the more serious critiques that specifically take aim at experiments looking at variation in the age-of-reproduction (like the work in Chapter 39). Promislow and Tatar (1998) note that in an iteroparous organism that is forced to reproduce only once early in life, selection on survival and reproduction later in life will be eliminated. Accordingly, they reasoned that any new mutation affecting survival or fertility after the age of reproduction would be effectively neutral and would over many generations rise to high frequency. Promislow and Tatar then make the rather precise prediction that "Given a per generation decline in fitness of between 0.1 and 1% due to mutation accumulation ... over a hundred or more generations one would expect to see a substantial decline in late-life fitness, perhaps as great as 50%" (Promislow and Tatar, 1998, pg. 307). Promislow and Tatar then suggest that populations forced to reproduce later in life are not improving their longevity but avoiding the deleterious effects of these late-acting mutations. Hence the laboratory experiment is erroneously suggesting that genetic variation to increase longevity exists in populations.

When replicate populations are maintained on this early reproduction schedule, then neutral alleles in each population will independently increase or decrease in each population due to random genetic drift. These late-acting alleles are neutral when there is only early reproduction although some may be deleterious when expressed in late-life. One way to detect the presence of late-acting deleterious alleles is through crosses of replicate populations to form hybrid populations. Two, independent, replicate populations are unlikely to have the same set of deleterious, late-acting alleles. Thus, a hybrid population formed from a cross of two replicate populations will produce individuals heterozygous for the deleterious alleles. Since these deleterious alleles are more likely to be recessive (see Borash et al., 2007) the hybrid populations should be cured of the deleterious effects.

This idea was put to a test by Borash et al. (2007) who studied five replicate, early reproducing populations that had been on this life cycle for 416 generations. These populations were kept at breeding sizes of greater than 1000 and effective sizes of 700–800 (Mueller et al., 2013). They found no evidence for the accumulation of deleterious mutations affecting longevity or female fecundity but some evidence for an effect on male virility.

Harshman and Hoffmann (2000) raise some related issues. They are concerned with some of the same issues as Promislow and Tatar (1998). However, they also discuss a new potential problem that long term selection can produce extreme and unusual changes in populations.

To support their claim about unusual and extreme outcomes of selection, Harshman and Hoffmann cite a paper reviewing more than 500 generations of selection for positive and negative geotaxis (Ricker and Hirsch, 1988). Specifically, Harshman and Hoffmann claim "Several responses to long-term selection might be rather specific manifestations of laboratory selection regimes. Apparently, long-term directional selection is capable of producing extreme changes or unusual changes in selected populations" (Harshman and Hoffmann, 2000, pg. 34). However, the populations used by Ricker and Hirsch have had a history of multiple extreme population bottlenecks including one generation of reproduction with one male and two females (Ricker and Hirsch, 1988). Even under better times these populations consisted of only 60 breeding pairs and these were separated into six groups of ten with the ten pairs with the highest or lowest geotaxis scores being put into one bottle, and so on. Thus, the opportunity for inbreeding, if there was between family variation in geotaxis, was increased by this protocol (Erlenmeyer-Kimling et al., 1962). Thus, it is not lab selection causing extreme outcomes but inbreeding, drift and population bottlenecks!

Field experiments

While controlled experiments can't be carried out in the field with the same precision as in a laboratory, certain types of experiments are possible. One of the most useful types of experiments involves manipulation of the environment or the phenotype of a organism in nature (Sinervo and Basolo, 1996). These manipulations may involve (1) changes to a single trait which does not affect other traits, (2) changes to a single trait which does affect other traits, (3) changes to resources that impact the phenotypes of the organism, and (4) alterations to the selective environment (Sinervo and Basolo, 1996). Manipulative experiments have been used in Chapters 31, 48, 59, and 62.

In Chapter 31, Pribil and Searcy (2001) manipulated the mating status of male birds and the quality of their territories. Thus, they used both (1) and (4) above. The selective environment (4) was manipulated in Chapters 48 and 59. In Chapter 48, Reznick et al. (1990) moved guppies to a new environment with a different predator. In Chapter 59, Schluter (1994) manipulated the competitive environment of threespine sticklebacks.

In Chapter 62, Dudley and Schmitt (1996) manipulated the light resources (3) and altered the density of plants (4).

GENETIC TECHNIQUES

Prior to the 1960's, studying genetic variation in natural populations was limited to genes which caused visible mutations. In Chapter 12 the allele frequencies of a wing color polymorphism were studied by Fisher and Ford (1947). By doing crosses between different wing color morphs, the genetic basis of this polymorphism was determined to be a single locus. However, the number of visible, single-locus polymorphims in natural populations is often extremely limited, making studies based on these genetic variants rare. In addition, for many organisms due to a lack of husbandry techniques, long developmental times, or other technical difficulties, it is not possible to do crosses to study the genetic basis of visible polymorphism.

However, this situation changed in 1966 with the publication by Lewontin and Hubby (1966). In this landmark paper they describe how protein polymorphisms could be studied in almost any organism allowing detailed studies of genetic variation in a large array of plants and animals. Over time, the molecular genetic techniques available for evolutionary studies has evolved. Below I review some of the techniques commonly used.

Protein polymorphisms

DNA may code for RNA's, serve a regulatory function, may have no known function, or may code for proteins. These proteins will often catalyze important physiological reactions. Tissue samples can be placed in a gel substrate (Fig. 5) to which an electric field is applied. The samples are prepared so they usually have a negative charge and are then pulled toward the positive pole. The distance moved from the cathode is a function of the size of the protein and its net charge. Thus, differences in the amino acid composition of allelic variants can be detected by this technique. For protein monomers the patterns of genetic variants are relatively straightforward (Fig. 5). See May (1998) for a discussion of more complicated patterns.

The widespread use of protein electrophoresis revealed much higher levels of genetic variation than most evolutionary biologists expected. This led to a debate about whether the genetic variation observed was maintained by natural selection or whether it was neutral (King and Jukes, 1969; Ayala,

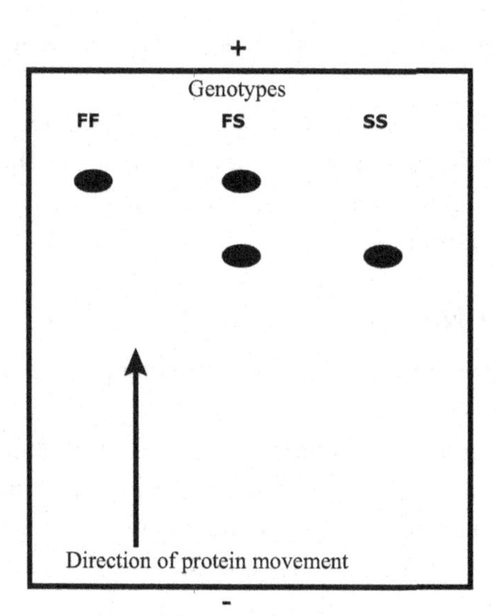

Figure 5 Protein electrophoresis. Samples are placed at the negative pole (cathode) of a gel substrate, usually starch or acrylamide. The proteins will usually have a net negative charge and thus be attracted to the anode. The distance the protein is pulled through the gel will be a function of the charge and size of the protein. The position of the protein is visualized by adding a staining solution specific to the enzyme. In this figure the protein is assumed to be a monomer and the pattern of three genotypes are shown, the fast (FF) and slow (SS) homozygote and the heterozygote (FS).

1976). One approach used in this debate was to find physiological differences between allelic variants that might translate into fitness differences (Koehn et al., 1980, 1988; Powers et al., 1979; Watt, 1985, 1986).

Quantitative genetics

Via and Lande (1985, Chapter 51) did much to stimulate interest in studying evolutionary questions in quantitative genetic terms. Grant (1986) used quantitative genetic estimates of the heritability of characters he believed are subject to selection (Chapter 53). There have been many other applications of quantitative genetic measurements to evolutionary ecology problems. A detailed application is described in Abrahamson and Weis (1997). There are a variety of ways to measure heritability, and genetic correlations [reviewed in Falconer (1989) and Roff (1997)]. A common problem that has dissuaded many from the widespread adoption of quantitative genetic measurements involves the large uncertainties associated with these estimates and the need to make measurements in the same environment where selection takes place in.

As techniques for the direct examination of DNA became available there has been a marked decline in the use of protein electrophoresis and to some extent quantitative genetics in evolutionary studies. I next review some of the DNA-based techniques.

Mitochondrial DNA

Mitochondrial DNA (mtDNA) is a small circular piece of DNA found in the mitochondria of plants and animals. In higher animals it is 16–20 kilobases long, there are 37 genes, no recombination and most genetic differences are single base pairs (Avise et al., 1987). It is also inherited matrilineally from the mitochondria in the mother's egg. In addition to these genes is the D-loop or control region which exerts control over replication and RNA transcription.

Since there is no recombination between mitochondrial DNA molecules, the genetic variants, or haplotypes, of mtDNA are like a single locus with many alleles. Haplotypes have been identified in two different ways, restriction site polymorphisms and direct sequencing. I review these techniques briefly.

Bacteria naturally have enzymes, called restriction endonucleases, which cut double stranded DNA at specific sites. These endonucleases protect the bacterium from foreign DNA. The bacteria's DNA is protected from these enzymes by methylation of their DNA. For instance, one restriction endonuclease isolated from the gut bacteria, *Escherichia coli*, and called *Eco*RI, cuts DNA when it encounters the sequence, 5′-GAATTC-3′. There are 100's of such enzymes that recognize specific, 4, 5, and 6 base pair sequences.

Isolated DNA from a sample can be treated with a specific restriction enzyme resulting in a mixture of DNA fragments of many different sizes. These DNA fragments can then be separated by size using electrophoresis, a process like that used in protein electrophoresis. The DNA in the gel can then be transferred to a nylon membrane in a process called Southern blotting. DNA fragments from a specific region of the genome can then be visualized by adding a solution with a DNA probe specific to the region of interest. That probe will have a radioactive or color marker that will allow visualization of the different DNA fragments from the genome region specific to the probe used (see Avise, 2004; chpt. 3 for details). By exposing the DNA to multiple restriction enzymes, a map of the location of the restriction sites can sometimes be generated. Additionally, with a sufficiently large sample the frequency of different restriction fragment length polymorphisms (RFLP) can be estimated.

The field of DNA sequencing is advancing so rapidly it is practically impossible to provide an up-to-date review which will not quickly become dated. However, most current techniques use variation of the original sequencing techniques developed by Maxam and Gilbert (1977, 1980) and Sanger et al. (1977). Each of these techniques is made easier by the polymerase chain reaction (PCR) technique (Mullis and Faloona, 1987). PCR has three steps. (1) A DNA sample is heated to denature the DNA and form single strands. (2) Primers that have unique sequences which flank the 3′ and 5′ end of the region of interest are allowed to anneal with the single stranded DNA. (3) Primer extension through the region of interest is carried out by a thermostable DNA polymerase (*Taq*). This process is then repeated 20 or more times to create a large sample of the genetic region of interest (see Avise, 2004; chpt. 3 for details).

The Sanger technique for DNA sequencing also starts out by denaturing double stranded DNA. A short DNA segment, called the primer, is allowed to anneal to the denatured DNA at a targeted location. DNA polymerase is then added with a mixture of the four deoxynucleotides and single dideoxynucleotide. This last molecule lacks the 3′OH group of a normal deoxynucleotides. Thus, when the cytosine dideoxynucleotide, for example, is added to the growing DNA strand instead of cytosine the DNA stops growing. Conditions are arranged so that the addition of the dideoxynucleotides is rare and random. These new fragments of DNA are labelled and can also be separated according to size by electrophoresis. After this is done the sequence can be determined by ordering the size fragments and determining which base pair stopped its growth. Thus, if the smallest fragment, which added only one base pair, is from the thymine dideoxynucleotide mixture then the first base pair in the sequence is thymine. The process continues by examining all fragments in order of size (see Avise, 2004; chpt. 3 for details). This technique has advanced considerably over the years (see for instance Bentley et al., 2008). The next generation high throughput methods are considerably more complicated but are also cheaper making large scale sequencing a common practice in evolutionary research.

Mitochondrial DNA has a high rate of evolution in animals and tends to show more variation among populations than within populations. This makes these markers useful for studying patterns of historical biogeography (Avise et al., 1987). However, the high rate of mutation for some mtDNA genes can make them problematic for more distantly related species (Dowling et al., 1996).

Genome sequencing

The ability to cheaply get genome sequence data has opened up a new era for experimental evolutionary ecology. While it is still too expensive to sequence very large numbers of individuals it is possible to get sequence information on a whole population. Several hundred individuals from a single population can be put in a pooled sample (pool-seq) and will yield the frequency of single nucleotide polymorphisms (SNP) across the genome. Thus, each population is a sample and pool-seq allows changes in SNP frequencies to be assessed over time. In combination with replicated experimental populations the technique is called evolve and resequence (E&R, Kofler and Schlötterer, 2013; Schlötterer et al., 2014). While still in its infancy E&R has been used to study evolution due to different ages-at-reproduction (Burke et al., 2010) and temperature adaptation (Mallard et al., 2018).

References

Abrahamson, W.G., Weis, A.E., 1997. Evolutionary Ecology Across Three Trophic Levels. Princeton Monographs in Population Biology. Princeton University Press, Princeton, NJ.

Avise, J.C., 2004. Molecular Markers, Natural History, and Evolution, second ed. Sinauer, Sunderland, MA.

Avise, J.C., Arnold, J., Ball, R.M., Bermingham, E., Lamb, T., Neigel, J.E., Reeb, C.A., Saunders, N.C., 1987. Intraspecific phylogeography: the mitochondrial DNA bridge between population genetics and systematics. Ann. Rev. Ecol. Syst. 18, 489–522.

Ayala, F.J. (Ed.), 1976. Molecular Evolution. Sinauer, Sunderland, MA.

Baldwin-Brown, J.G., Long, A.D., Thornton, K.R., 2014. The power to detect quantitative trait loci using resequenced, experimentally evolved populations of diploid, sexual organisms. Mol. Biol. Evol. 31, 1040–1055.

Bentley, D.R., et al., 2008. Accurate whole human genome sequencing using reversible terminator chemistry. Nature 456, 53–59.

Borash, D.J., Rose, M.R., Mueller, L.D., 2007. Mutation accumulation affects male virility in *Drosophila* selected for later reproduction. Physiol. Biochem. Zool. 80, 461–472.

Britnacher, J.G., 1981. Genetic variation and genetic load due to the male reproductive component of fitness in *Drosophila*. Genetics 97, 719–730.

Burke, M.K., Dunham, J.P., Shahrestani, P., Thornton, K.R., Rose, M.R., Long, A.D., 2010. Genomewide analysis of a long-term evolution experiment with *Drosophila*. Nature 467, 587–590.

Dennis, B., Costantino, R.F., 1988. Analysis of steady-state populations with the Gamma abundance model and its application to *Tribolium*. Ecology 69, 1200–1213.

Dennis, B., Desharnais, R.A., Cushing, J.M., Costantino, R.F., 1995. Nonlinear demographic dynamics: mathematical models, statistical methods, and biological experiments. Ecol. Monogr. 65, 261–281.

Desharnais, R.A., Costantino, R.F., 1982. The approach to equilibrium and the steady state probability distribution of adult numbers in *Tribolium brevicornis*. Am. Nat. 119, 102–111.

Dobzhansky, T., Spassky, B., Tidwell, T., 1963. Genetics of natural populations. XXXII. Inbreeding and mutational loads in natural populations of *Drosophila pseudoobscura*. Genetics 48, 361–373.

Dowling, T.E., Moritz, C., Palmer, J.D., Rieseberg, L.H., 1996. Nucleic acids III: analysis of fragments and restriction sites. In: Hillis, D.M., Moritz, C., Mable, B.K. (Eds.), Molecular Systematics, second ed. Sinauer, Sunderland, MA.

Dudley, S.A., Schmitt, J., 1996. Testing the adaptive plasticity hypothesis: density-dependent selection on manipulated stem length in *Impatiens capensis*. Am. Nat. 147, 445–465.

Erlenmeyer-Kimling, L., Hirsch, J., Weiss, J.M., 1962. Studies in experimental behavioral genetics III. Selection and hybridization analyses of individual differences in the sign of geotaxis. J. Comp. Physiol. Psych. 55, 722–731.

Falconer, D.S., 1989. Introduction to Quantitative Genetics. Longmans, New York.

Feldman, M.W., Franklin, I.R., Thomson, G., 1974. Selection in complex genetic systems: I. The symmetric equilibrium of the three locus symmetric viability model. Genetics 76, 135–162.

Fisher, R.A., Ford, E.B., 1947. The spread of a gene in natural conditions in a colony of the moth *Panaxia dominula* L. Heredity 1, 143–174.

Gewin, V., 2013. Not all species deteriorate with age. Nature. https://doi.org/10.1038/nature.2013.14322.

Grant, P.R., 1986. Ecology and Evolution of Darwin's Finches. Princeton University Press, Princeton, NJ.

Harshman, L.G., Hoffmann, A.A., 2000. Laboratory selection experiments using *Drosophila*: what do they really tell us? Trends Ecol. Evol. 15, 32–36.

Jones, O.R., Scheuerlein, A., Salguero-Gomez, R., Giovanni Camarda, C., Schaible, R., Casper, B.B., Dahlgren, J.P., Ehrlen, J., Garcõa, M.B., Menges, E.S., Quintana-Ascencio, P.F., Caswell, H., Baudisch, A., Vaupel, J.W., 2014. Diversity of ageing across the tree of life. Nature 505, 169–173.

Karlin, S., 1975. General two-locus selection models: some objectives, results and interpretations. Theor. Popul. Biol. 7, 364–398.

Keddy, P.A., 2001. Competition, second ed. Dordrect, Boston.

King, J.L., Jukes, T.H., 1969. Non-Darwinian evolution: random fixation of selectively neutral mutations. Science 164, 788–798.

Koehn, R.K., Newell, R.I.E., Immermann, F., 1980. Maintenance of an aminopeptidase allele frequency cline by natural selection. Proc. Natl. Acad. Sci. U.S.A. 77, 5385–5389.

Koehn, R.K., Diehl, W.J., Scott, T.M., 1988. The differential contribution of individual enzymes of glycolysis and protein catabolism to the relationship between heterozygosity and growth rate in the coot clam, *Mulinia lateralis*. Genetics 118, 121–130.

Kofler, R., Schlötterer, C., 2013. A guide for the design of evolve and resequencing studies. Mol. Biol. Evol. 31, 474–483.

Lewontin, R.C., Hubby, J.L., 1966. A molecular genetic approach to the study of genic heterozygosity in natural populations. II amounts of variation and degree of heterozygosity in natural populations of *Drosophila pseudoobscura*. Genetics 54, 595–609.

Mallard, F., Nolte, V., Tobler, R., Kapun, M., Schlötterer, C., 2018. A simple genetic basis of adaptation to a novel thermal environment result in complex metabolic rewiring in *Drosophila*. Genome Biol. 9, 119–133.
Marinkovic, D., 1967. Genetic loads affecting fecundity in natural populations of *Drosophila pseudoobscura*. Genetics 56, 61–71.
Maxam, A.M., Gilbert, W., 1977. A new method for sequencing DNA. Proc. Natl. Acad. Sci. U.S.A. 74, 560–564.
Maxam, A.M., Gilbert, W., 1980. Sequencing end-labelled DNA with base-specific chemical cleavages. Methods Enzymol. 65, 499–559.
May, B., 1998. Starch gel electrophoresis of allozymes. In: Hoelzel, A.R. (Ed.), Molecular Genetic Analysis of Populations. IRL Press, Oxford.
Mueller, L.D., 1985. The evolutionary ecology of *Drosophila*. Evol. Biol. 19, 37–98.
Mueller, L.D., 1987. Evolution of accelerated senescence in laboratory populations of *Drosophila*. Proc. Natl. Acad. Sci. U.S.A. 84, 1974–1977.
Mueller, L.D., 1988. Density-dependent population growth and natural selection in food limited environments: the *Drosophila* model. Am. Nat. 132, 786–809.
Mueller, L.D., 1997. Theoretical and empirical examination of density-dependent selection. Annu. Rev. Ecol. Evol. Syst. 28, 269–288.
Mueller, L.D., Graves Jr., J.L., Rose, M.R., 1993. Interactions between density-dependent and age-specific selection in *Drosophila melanogaster*. Funct. Ecol. 7, 469–479.
Mueller, L.D., Joshi, A., 2000. Stability in Model Populations. Monographs in Population Biology. Princeton University Press, Princeton, NJ.
Mueller, L.D., Joshi, A., Santos, M., Rose, M.R., 2013. Effective population size and evolutionary dynamics in outbred laboratory populations of *Drosophila*. J. Genet. 92, 349–361.
Mueller, L.D., Phillips, M.A., Barter, T.T., Greenspan, Z., Rose, M.R., 2018. Genome-wide mapping of gene-phenotype relationships in experimentally evolved populations. Mol. Biol. Evol. 35, 2085–2095.
Mullis, K., Faloona, F., 1987. Specific synthesis of DNA in vitro via a polymerase catalyzed chain reaction. Methods Enzymol. 155, 335–350.
Oster, G.W., Wilson, E.O., 1978. Social Insects. Monographs in Population Biology 12. Princeton Univ. Press, Princeton, NJ.
Pearl, R., 1927. The growth of populations. Q. Rev. Biol. 2, 532–548.
Peters, C.S., Mangel, M., Costantino, R.F., 1989. Stationary distribution of population size in *Tribolium*. Bull. Math. Biol. 51, 625–638.
Powers, D.A., Greaney, G.S., Place, A.R., 1979. Physiological correlation between lactate dehydrogenase genotype and haemoglobin function in killifish. Nature 277, 240–241.
Pribil, S., Searcy, W.A., 2001. Experimental confirmation of the polygyny threshold model for red-winged blackbirds. Proc. R. Soc. Lond. B 268, 1643–1646.
Promislow, D.E.L., Tatar, M., 1998. Mutation and senescence: where genetics and demography meet. Genetica 102/103, 299–314.
Prout, T., 1980. Some relationships between density-independent selection and density dependent population growth. Evol. Biol. 13, 1–68.
Reznick, D., Bryga, H., Endler, J.A., 1990. Experimentally induced life-history evolution in a natural population. Nature 346, 357–359.
Ricker, J.P., Hirsch, J., 1988. Reversal of genetic homeostasis in laboratory populations of *Drosophila melanogaster* under long-term selection for geotaxis and estimates of gene correlates: evolution of behavior-genetic systems. J. Comp. Psychol. 102, 203–214.
Roff, D.A., 1997. Evolutionary Quantitative Genetics. Chapman and Hall, New York.
Rose, M.R., Nusbaum, T.J., Chippindale, A.K., 1996. Laboratory evolution: the experimental wonderland and the Cheshire cat. In: Rose, M.R., Lauder, G.V. (Eds.), Adaptation. Academic Press, San Diego.

Roughgarden, J., 1976. Resource partitioning among competing species-a coevolutionary approach. Theor. Popul. Biol. 9, 388–424.

Stephens, D.W., Krebs, J.R., 1986. Foraging Theory. Princeton University Press.

Sanger, F., Nicklen, S., Coulsen, A.R., 1977. DNA sequencing with chain-terminating inhibitors. Proc. Natl. Acad. Sci. U.S.A. 74, 5463–5467.

Schlötterer, C., Kofler, R., Versace, E., Tobler, R., Franssen, S.U., 2014. Combining experimental evolution with next-generation sequencing: a powerful tool to study adaptation from standing genetic variation. Heredity 114, 431–440.

Schluter, D., 1994. Experimental evidence that competition promotes divergence in adaptive radiation. Science 266, 798–801.

Sinervo, B., Basolo, A.L., 1996. Testing adaptation using phenotypic manipulations. In: Rose, M.R., Lauder, G.V. (Eds.), Adaptation. Academic Press, San Diego.

Via, S., Lande, R., 1985. Genotype-environment interaction and the evolution of phenotypic plasticity. Evolution 39, 505–522.

Watt, W.B., 1985. Bioenergetics and evolutionary genetics: opportunities for new synthesis. Am. Nat. 125, 118–143.

Watt, W.B., 1986. Power and efficiency as indices of fitness in metabolic organization. Am. Nat. 127, 629–653.

Wright, S., 1982. The shifting balance theory of macroevolution. Ann. Rev. Genet. 16, 1–19.

Index

'*Note*: Page numbers followed by "f" indicate figures, "t" indicate tables and "b" indicate boxes.'

N

O

P

Q

R

S

T

U

V

W

Printed in the United States
By Bookmasters